Claudine Groß

A soft-tetramer model for diblock copolymer melts

Claudine Groß

A soft-tetramer model for diblock copolymer melts

Structure formation and geometric characterization

Südwestdeutscher Verlag für Hochschulschriften

Imprint
Any brand names and product names mentioned in this book are subject to trademark, brand or patent protection and are trademarks or registered trademarks of their respective holders. The use of brand names, product names, common names, trade names, product descriptions etc. even without a particular marking in this work is in no way to be construed to mean that such names may be regarded as unrestricted in respect of trademark and brand protection legislation and could thus be used by anyone.

Publisher:
Südwestdeutscher Verlag für Hochschulschriften
is a trademark of
Dodo Books Indian Ocean Ltd., member of the OmniScriptum S.R.L Publishing group
str. A.Russo 15, of. 61, Chisinau-2068, Republic of Moldova Europe
Printed at: see last page
ISBN: 978-3-8381-2449-0

Zugl. / Approved by: Mainz, Johannes Gutenberg-Universität, Diss., 2010

Copyright © Claudine Groß
Copyright © 2011 Dodo Books Indian Ocean Ltd., member of the OmniScriptum S.R.L Publishing group

Abstract

The ability of block copolymers to spontaneously self-assemble into a variety of ordered nano-structures not only makes them a scientifically interesting system for the investigation of order-disorder phase transitions, but also offers a wide range of nano-technological applications. The architecture of a diblock is the most simple among the block copolymer systems, hence it is often used as a model system in both experiment and theory.

We introduce a new soft-tetramer model for efficient computer simulations of diblock copolymer melts. The instantaneous non-spherical shape of polymer chains in molten state is incorporated by modeling each of the two blocks as two soft spheres. The interactions between the spheres are modeled in a way that the diblock melt tends to microphase separate with decreasing temperature.

Using Monte Carlo simulations, we determine the equilibrium structures at variable values of the two relevant control parameters, the diblock composition and the incompatibility of unlike components. The simplicity of the model allows us to scan the control parameter space in a completeness that has not been reached in previous molecular simulations. The resulting phase diagram shows clear similarities with the phase diagram found in experiments. Moreover, we show that structural details of block copolymer chains can be reproduced by our simple model.

We develop a novel method for the identification of the observed diblock copolymer mesophases that formalizes the usual approach of direct visual observation, using the characteristic geometry of the structures. A cluster analysis algorithm is used to determine clusters of each component of the diblock, and the number and shape of the clusters can be used to determine the mesophase.

We also employ methods from integral geometry for the identification of mesophases and compare their usefulness to the cluster analysis approach.

To probe the properties of our model in confinement, we perform molecular dynamics simulations of atomistic polyethylene melts confined between graphite surfaces. The results from these simulations are used as an input for an iterative coarse-graining procedure that yields a surface interaction potential for the soft-tetramer model.

Using the interaction potential derived in that way, we perform an initial study on the behavior of the soft-tetramer model in confinement. Comparing with experimental studies, we find that our model can reflect basic features of confined diblock copolymer melts.

Zusammenfassung

Die Fähigkeit von Block-Copolymeren, sich spontan in eine Vielzahl von geordneten Nanostrukturen zu organisieren, macht sie nicht nur zu einem wissenschaftlich interessanten System für die Untersuchung von Ordnungs-Unordnungs-Phasenübergängen, sondern bietet auch eine breite Palette von nanotechnologischen Anwendungen. Das Diblock-System ist das einfachste unter den Block-Copolymeren, daher wird es sowohl im Experiment als auch in der Theorie oft als Modellsystem verwendet.

Wir stellen ein neues Tetramer-Modell für effiziente Computersimulationen von Diblock-Copolymerschmelzen vor. Die instantane nichtkugelförmige Gestalt der Polymerketten in der Schmelze wird dabei durch die Modellierung jedes der beiden Blöcke als zwei weiche Kugeln berücksichtigt. Die Wechselwirkungen zwischen den Kugeln werden dabei so gewählt, dass die Diblock-Schmelze bei abnehmender Temperatur zur Mikrophasenseparation neigt.

Mittels Monte-Carlo-Simulationen bestimmen wir die Gleichgewichtsstrukturen für verschiedene Werte der beiden relevanten Kontrollparameter, der Zusammensetzung des Diblocks und der Unvereinbarkeit der beiden gegensätzlichen Komponenten. Die Einfachheit des Modells erlaubt es, den Raum der Kontrollparameter in einer Gründlichkeit zu untersuchen, die bisher in molekularen Simulationen unerreicht ist. Das daraus resultierende Phasendiagramm zeigt deutliche Ähnlichkeiten mit dem Phasendiagramm, das in Experimenten bestimmt wurde. Darüber hinaus zeigen wir, dass unser einfaches Modell strukturelle Eigenschaften von Block-Copolymer-Ketten reproduzieren kann.

Wir entwickeln eine neuartige Methode zur Identifizierung der beobachteten Mesophasen, die mit Hilfe der charakteristischen Geometrie der Strukturen die übliche Vorgehensweise der direkten Betrachtung formalisiert. Ein Cluster-Analyse-Algorithmus wird verwendet, um Cluster bestehend aus einer Diblock-Komponente zu bestimmen, und die Anzahl und die Form der Cluster kann verwendet werden, um die Mesophase zu bestimmen.

Wir setzen auch Methoden aus der Integralgeometrie für die Identifizierung von Mesophasen ein und vergleichen ihren Nutzen mit dem der Cluster-Analyse-Methode.

Um die Eigenschaften unseres Modells zu untersuchen, wenn es zwischen Oberflächen eingeschlossen ist, führen wir Molekulardynamik-Simulationen von atomistischen Polyethylenschmelzen zwischen Graphitoberflächen durch. Die Ergebnisse dieser Simulationen werden als Ausgangspunkt für einen iterativen Prozess verwendet, der ein Potential für die Oberflächenwechselwirkung im Tetramer-Modell erzeugt.

Mithilfe des so abgeleiteten Wechselwirkungspotentials führen wir eine erste Studie zum

Verhalten des Modells im Einschluss zwischen Oberflächen durch. Im Vergleich mit experimentellen Untersuchungen finden wir, dass unser Modell grundlegende Eigenschaften von eingeschlossenen Diblock-Copolymerschmelzen wiedergeben kann.

초록

공중합체(block copolymer)는 스스로 자기조립하여 다양한 질서를 지닌 나노 구조를 가지기 때문에 질서-무질서 상전이 연구에 적합할 뿐만 아니라 나노기술로 폭넓게 응용된다. 그 중에서 이중 공중합체(diblock copolymer)는 가장 간단한 구조를 가지고 있으므로, 실험이나 이론적 연구에 적합한 모델로 활용되고 있다.

본 연구에서는 이중 공중합체 멜트에 대한 컴퓨터 시뮬레이션 연구에 효율적인 새로운 soft-tetramer 모델을 도입하였다. 용융 상태에서 일시적으로 비구형(nonspherical) 모양인 고분자 사슬은 두 개의 무른 구체를 지닌 두 블록으로 정의되었으며, 구체간의 상호작용은 온도가 내려갈 때 이중 공중합체가 마이크로상으로 분리될 수 있도록 고안되었다.

또한 매개 변수로 두 블록의 조성(composition)과 서로 다른 성분간의 불화합성(incompatibility) 개념을 도입한 몬테칼로 시뮬레이션을 이용하여 멜트의 평형 구조를 계산하였다. 본 모델은 매우 간단한 개념을 사용하였기 때문에 기존 연구에서는 불가능하였던 모든 매개 변수 영역에서의 결과를 얻을 수 있었다. 뿐만 아니라 시뮬레이션 결과 얻어진 상그림(phase diagram)은 실험을 통해 알려진 것과 매우 유사하였을 뿐만 아니라 이중 공중합체의 자세한 구조적 특성까지 도출할 수 있었다.

한편 시뮬레이션 연구에서 발견된 이중 공중합체의 준결정상(mesophase)을 식별하기 위한 분석법을 고안하였는데, 구조의 기하학적 특성을 이용하여 육안으로 식별하는 방법을 수식화하였다. 클러스터 알고리듬 분석법을 이용하여 먼저 이중 공중합체의 각 성분들의 클러스터를 결정한 후, 클러스터의 모양과 개수를 사용하여 준결정상을 결정하였다.

또한 적분 기하학을 통한 준결정상 분석법을 도입하였으며 클러스터 분석 방법을 사용한 결과와 비교하였다.

그리고 공간적으로 가두어진(confined) 고분자 계로 본 모델을 확장하기 위하여 흑연 표면 사이에 가두어진 폴리에텔렌 멜트상의 분자 동역학 모의실험을 수행하였다. 분자 동역학 모의실험으로 얻어진 결과는 대충갈기(coarse graining) 과정을 통해 soft-tetramer 모델에 필요한 표면 상호작용 퍼텐셜로 변환되었다.

얻어진 표면 상호작용 퍼텐셜을 적용하여 가두어진 계에서도 soft-tetramer 모델이 잘 작동하는지 알아보기 위한 기초적인 연구를 수행하였다. 실험결과와 비교하였을 때, 본 모델은 가두어진 이중 공중합체 멜트의 기본적인 특성을 잘 구현할 수 있는 것으로 나타났다.

Contents

1. **Motivation**   1

2. **General concepts**   5
   - 2.1. Introduction to polymers 5
     - 2.1.1. Theoretical concepts 5
     - 2.1.2. Coarse graining 7
   - 2.2. Diblock copolymer melts 9
     - 2.2.1. Experimental methods 11
     - 2.2.2. Theoretical concepts 12
     - 2.2.3. Phase diagram 14

3. **Model**   19
   - 3.1. Introduction 19
   - 3.2. Model Description 21
     - 3.2.1. Interaction potentials 21
     - 3.2.2. Control parameters 22
   - 3.3. Invariant degree of polymerization 24

4. **Geometric characterization of mesophases**   27
   - 4.1. Mesophase identification by cluster analysis 27
     - 4.1.1. Cluster identification algorithm 28

		4.1.2. Mesophase identification	32

        4.1.2. Mesophase identification 32
        4.1.3. Shape parameters 36
    4.2. Minkowski functionals 37
        4.2.1. Integral geometry 37
        4.2.2. Mapping and calculation 38
        4.2.3. Minkowski functionals for diblock copolymer mesophases 40

5. Bulk simulations   45

    5.1. Simulation method 45
        5.1.1. The Monte Carlo Method in the canonical ensemble 45
        5.1.2. Simulation details 48
    5.2. Phase diagram 49
        5.2.1. Examples of characterization 49
        5.2.2. Phase diagram: results and discussion 58
    5.3. Cluster algorithm on discretized configurations 62
    5.4. Shape parameters 65
    5.5. Euler characteristic 66
    5.6. Structural details 67
        5.6.1. Chain stretching 67
        5.6.2. Saupe order tensor 69
        5.6.3. Lattice site correlation order parameter ψ 73

6. Confined polymer melts   75

    6.1. MD Simulations of confined polyethylene melts 75
        6.1.1. Simulation details – polyethylene 76
        6.1.2. Confined polyethylene chains 80
    6.2. Derivation of the effective potential 83
        6.2.1. Iterative Boltzmann Inversion 83
        6.2.2. Iteration procedure 84
        6.2.3. Root mean square deviation 87
        6.2.4. Effective potential 88
    6.3. Soft-tetramer model in confinement 90
        6.3.1. Preferential walls 91
        6.3.2. Neutral walls 97
        6.3.3. Discussion 98

7. Conclusions	101
A. Cluster analysis for disordered states	105
B. Grand-canonical Monte Carlo moves	109
B.1. Simulation technique	109
C. The molecular dynamics method	113

List of Figures

1.1. Schematics of a diblock copolymer nanocomposite 2
2.1. Architecture of a diblock copolymer chain. Mechanism of microphase separation for the lamellar mesophase. 10
2.2. Schematic representation of common diblock copolymer mesophases 14
2.3. Comparison of the experimental phase diagram with the phase diagram from self-consistent field theory 16
2.4. Partial phase diagram using lattice-based Monte-Carlo simulations 18

3.1. Bonded and non-bonded interaction potentials. The inset shows a schematic representation of the model (not to scale). 23

4.1. Example simulation snapshot, lamellar mesophase. 28
4.2. Network example consisting of seven nodes of the same species. 30
4.3. Simulation snapshot of the cylinder phase at $\rho_c = 1.3$, $f = 0.78$ and $\chi N = 42$ ((a) and (b)) and of the perforated lamella phase at $\rho_c = 1.3$, $f = 0.72$ and $\chi N = 21$ (c). 32
4.4. Comparison for choices (1.)–(4.) of next neighbour radii in the cylindrical phase, parameters $f = 0.78, \chi N = 42, \rho_c = 1.3$. 33
4.5. Recomparison for choices (1.)–(2.) of next neighbor radii in the perforated lamellar phase, parameters $f = 0.75, \chi N = 21, \rho_c = 1.3$. 34

List of Figures

4.6. Location of ideal solids in the (K_1, K_2)-plane. 37
4.7. Mapping schematics for a simple two-dimensional two-sphere example. . . . 39
4.8. Surface vs. discretization for ideal cylinders. 41

5.1. Lamellar (L) phase of the soft-tetramer model observed at $f = 0.66$ and $\chi N = 30$ ($L = 11$, $\rho_c = 1$). 50
5.2. Cluster number distributions for majority (A) and minority (B) phases for a lamellar mesophase for $\rho_c = 1.3$, $f = 0.66$ and $\chi N = 30$ (a), and distributions for the three mean eigenvalues of the gyration tensor of the clusters in (b). 51
5.3. Snapshot of the perforated lamella (PL) phase at $\rho_c = 1.3$, $f = 0.72$ and $\chi N = 21$ and section showing just one perforated minority lamella. 53
5.4. Cluster number distributions for majority (A) and minority (B) components for the perforated lamella mesophase at $\rho_c = 1.3$, $f = 0.72$ and $\chi N = 21$ (a) and distributions for the three eigenvalues of the gyration tensor of the clusters (b). 54
5.5. Simulation snapshot of the cylinder phase at $\rho_c = 1.3$, $f = 0.81$ and $\chi N = 48$. 56
5.6. Cluster number distributions for majority (A) and minority (B) phases for the cylinder mesophase at $\rho_c = 1.3$ (a) and distributions for the three eigenvalues of the gyration tensor of the clusters (b). 57
5.7. Phase diagram of the soft-tetramer model for box size $L = 12$, chain number density $\rho_c = 1$. 59
5.8. Phase diagram of the soft-tetramer model for box size $L = 11$, chain number density $\rho_c = 1.3$. 60
5.9. Dependence of the location of the disorder-order transition on box length and density, shown examplarily for the shape parameters K_1 and K_2. 63
5.10. Results of the cluster algorithm applied on discretized configurations compared with cluster algorithm applied directly to configuration for L=11, f=0.66, $\chi N = 30$, $\rho = 1.3$. 64
5.11. Composition dependence of the shape parameters K_1 and K_2 for $\chi N = 33$ and $\chi N = 42$ in the regime of the ordered mesophases, $\rho_c = 1.3$. 65
5.12. Location of the simulated lamellar (L) and cylindrical (C) structures in the (K_1, K_2) plane. 66
5.13. Composition dependence of the Euler characteristic χ_e of the minority component 68

List of Figures

5.14. Distance between the centers of mass of the A and B blocks as a function of χN at $f = 0.5$, normalized to the limit in the disordered state. 70
5.15. Molecule directors attached to the tetramer. 71
5.16. χN dependence of the largest eigenvalue of the Saupe order tensor for all investigated molecule directors, for both systems at symmetric composition $f = 0.5$. 72
5.17. Correlation order parameter ψ. 74

6.1. Schematics of the united atom model for alkanes 76
6.2. Lennard-Jones potential. 77
6.3. Principle of angle bending. 78
6.4. Principle of dihedral angle. 78
6.5. Ryckaert-Bellemans potential. 79
6.6. Simulation snapshot of the confined polyethylene melt 79
6.7. Density profiles for half chains and monomers (C100). 81
6.8. Simulation snapshot of two C100 chains, each with one half adsorbed at the surface. 82
6.9. Probability distribution profile of centers of mass of C300 half chains, distribution averaged over samples from the last 400 µs, and monomer density profile. 83
6.10. C300 whole-chain center of mass density profile. 84
6.11. Distribution from MD simulation (Figure 6.9), symmetrized, mapped to the soft-tetramer model length scale and smoothed by a five-point running average. 85
6.12. Results from the first iteration 87
6.13. Root mean square deviation. 88
6.14. Results from the final iteration 89
6.15. Effective potential used for attractive interaction ($r_{\min} \approx 0.7, \epsilon \approx 0.5$) and the same potential shifted and cut off at the minimum for purely repulsive interaction. 90
6.16. Distribution of $\langle G_3^{AB} \rangle$ resulting from the application of the cluster algorithm to discretized configurations for the bulk system $f = 0.5$, $\chi N = 24$, $\rho_c = 1$. 91
6.17. A and B sphere distribution profiles for various surface distances L_z. ... 93
6.18. A and B sphere distribution profiles for various surface distances L_z. ... 94
6.19. Simulation snapshots for selected surface distances L_z. 95
6.20. The soft-tetramer model confined between neutral walls 97

A.1. Distributions of cluster numbers and gyration tensor eigenvalues in the disordered phase. $f = 0.5$, $\chi N = 1$, $\rho_c = 1.3$, $L = 11$. 105
A.2. Simulation snapshot at parameters $f = 0.5$, $\chi N = 1$, $\rho_c = 1.3$, $L = 11$. . . . 106
A.3. Distributions of cluster numbers and gyration tensor eigenvalues in the disordered phase. $f = 0.5$, $\chi N = 6$, $\rho_c = 1.3$, $L = 11$. 107
A.4. Distributions of cluster numbers and gyration tensor eigenvalues in the disordered phase. $f = 0.81$, $\chi N = 21$, $\rho_c = 1.3$, $L = 11$. 107
A.5. Simulation snapshot at parameters $f = 0.81$, $\chi N = 21$, $\rho_c = 1.3$, $L = 11$. . 108

CHAPTER 1

Motivation

Diblock copolymers are a system of great interest due to their wide range of applications in nanotechnology. Polymeric materials hold considerable potential for many technological applications because of their low cost, mechanic flexibility and low weight. The performance of organic opto-electronic devices in the field of photovoltaics and light emitting diodes depends crucially on understanding and controlling structures on the nanometer scale [1]. The fact that block copolymers can self-assemble into a variety of ordered microstructures on the nanometer length scale which *e.g.* happens to be also the characteristic length scale of exciton diffusion makes them extremely useful for those applications.

A photovoltaic cell, for instance, generally consists of an electron donating component in which light can be absorbed exciting an electron, and a component that accepts the excited electron from the donor. When light is absorbed in the donor material, the excited electron forms an *exciton*, *i.e.* a quasi-particle consisting of an electron and an electron hole.

After photo-excitation and charge separation, the electron and the hole are then transported back to the appropriate electrodes through the accepting and donating domains, respectively. This process works best if the distance between donor-acceptor interfaces is in the order of the diffusion length of the exciton, so that most excitons are able to reach an interface prior to recombination. Moreover, the donor and acceptor phases have to form

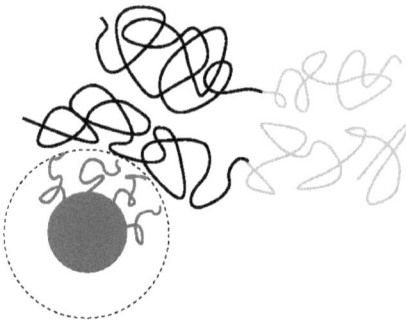

Figure 1.1.: Schematics of a diblock copolymer nanocomposite. Due to the typically unfavorable interactions between the polymer and the particle surface, most particles have to be pretreated to tailor their surface for better compatibility, *e.g.* by attaching ligands to the surface.

continuous pathways to the electrodes to make efficient charge transport possible.
The patterning needed for this purpose is achieved easier and cheaper by self-assembly of block copolymers than via conventional lithographic patterning techniques and the morphology is easier to control than it is the case for bulk hetero-junctions obtained from quenched spinodal decomposition of binary polymer blends. Moreover, because the self-assembly results in structures that are in thermodynamic equilibrium, the devices are less subject to aging.
The situation is similar for organic light emitting diodes; the transport of electrons and holes has to be balanced, which can be achieved by pattern optimization.
For these reasons, the understanding of block copolymer self-assembly to materials that are appropriate for organic electronics is subject of great interest.

The properties of self-assembled materials can be enhanced by combination with nanoparticles with a size that is comparable to the size of a polymer coil, *i.e.* by forming so-called nanocomposites, and also materials with new properties can be obtained. A sketch of a nanocomposite is depicted in Figure 1.1. Incorporation of nanoparticles in a block copolymer matrix allows control over the spatial distribution and orientation of the nanosized filler materials according to the desired application. In this way it is possible to transfer the long range order from the block copolymer matrix to the nanoparticles, thus using the copolymer as a scaffold [2]. On the other hand, the structure and orientation of the matrix can also be changed by the nanoparticles [3].

There is a vast variety of applications for block copolymer based nanocomposites, in optoelectronics, for instance, it is possible to enhance charge transport by inserting carbon nanoparticles [4]. Another application example are lithium batteries with electrolytes based on block copolymers, where one block is made up of a lithium-salt-solvating component. When doped with a lithium salt, the material becomes an electrolyte that is able to conduct lithium ions [5]. Block copolymer based nanocomposite thin films have also been used in flash memory applications to improve the robustness of the devices [6], and these are just a few examples from a list of numerous possible applications.

Since the properties of the nanocomposites highly depend on the microstructure, it is indispensable to gain a precise knowledge of the mechanisms of structure formation. However, a proper understanding is still missing [2].

The aim of this work is to pave the way towards a deeper understanding of the structure formation in block copolymers and block copolymer based nanocomposites by providing a model that allows efficient computer simulations by modeling the diblock copolymer molecules on the relevant scale, and that permits modeling of nanoparticles and copolymer molecules on the same scale.

CHAPTER 2

General concepts

2.1. Introduction to polymers

The term *polymer* originates from the Greek words πολύ, meaning "many", and μέρος, meaning "part". A polymer is a molecule constructed by covalently linking (*polymerizing*) repeat units (the so-called *monomers*) to form a long chain. If the monomers are identical, the molecule is called a homopolymer, whereas it is referred to as a copolymer if it consists of at least two different types of chemically distinct monomers. If these different types are grouped together in long sequences (or "blocks") of homopolymers, they form so-called *block copolymers*, which will be introduced in more detail in section 2.2.

2.1.1. Theoretical concepts

Ideal Chains

The most basic model in polymer physics is that of the ideal chain. It neglects volume interactions between monomers and describes the polymer chain as a random walk in space consisting of uncorrelated steps according to a Gaussian distribution. For this reason, it is also referred to as *Gaussian chain*. Despite its simplicity, it gives elucidating insight into the physics of polymers, because it provides a good approximation of the behavior of polymers in melts and certain solvents.

A very simple instance of ideal chain model is the freely jointed chain, by means of which section 2.1.2 will show that the size of the ideal chain is proportional to N^ν, with the number of segments N and the exponent $\nu = 1/2$ for the Gaussian chain.

Real Chains

In real chains, chain segments that are separated along the chain will interact if they come close enough. Hence the description by a random walk where the arrangement of chain segments is completely uncorrelated is not justified in the case of a real chain. To introduce volume interactions into the description of an ideal chain, the random walk is replaced by a self-avoiding walk, where a segment takes a certain volume that can not be occupied by another segment. Compared to ideal chains, the average size is larger: the so-called *excluded volume effect* gives rise to an increase in size from $\nu = 1/2$ for the Gaussian chain to $\nu \approx 3/5$, or more exactly $\nu \approx 0.588$, as has been found in simulations and renormalization group calculations for self-avoiding chains [7, 8].

Polymer Melts

The extreme case of a concentrated solution is called a "melt", *i.e.* a liquid that contains no solvent and is composed only of polymer chains. A particular feature of polymer chains in melts or in concentrated solutions is that their confirmation can be approximately described on large scales by an ideal Gaussian chain [7].

We can sense the physical meaning of this phenomenon by considering a single test chain: The excluded volume effect leads to a repulsive potential generated by the monomers of the test chain that would cause an expansion of the coil, if it was not canceled out by a counteracting potential generated by the identical molecules that surround the test chain in a homogeneous melt. Hence, the chain is subjected to no net force and remains unperturbed, and Gaussian statistics are satisfied. This oversimplified argument is, however, not completely correct and subtle long-range correlations along the chain persist in the melt [9].

2.1.2. Coarse graining

Since polymers exhibit structures on a wide range of length scales, from bond lengths in the order of angstroms to the typical extension of a molecule in the order of several nanometers or macroscopic properties occurring in the order of millimeters, one can not describe all of these phenomena by atomistic models that include all chemical details. The advantage of high detail is offset by the disadvantage that one is limited to systems of few molecules and short time scales.

Hence, polymeric materials have to be described on different scales, adapted to the respective problem that is to be addressed, by neglecting degrees of freedom that are irrelevant on that scale. On the microscale, relatively realistic models that include chemical details describe polymer chains at an atomistic level, on the mesoscale, simplified *coarse-grained* molecule models are used, where several real monomers are represented by simplified segments (see below), while on the macroscale, the materials are described by continuous fields.

While the chemistry of the polymer depends on atomistic details like the electron structure and the bonding between the monomers, there are many macroscopic properties f for which these chemical details are irrelevant. As a consequence, all chains behave in the same way when they are probed at a higher length scale, *i.e.* a certain universality appears, and this behavior can be described by so-called *scaling laws*, $f \propto N^x$. The exponent x is universal and independent of the microscopic details. A typical example of a physical quantity that shows this universal behavior is the size of a molecule, given by the expectation value of the distance between the two ends of the chain. Its scaling law has already been introduced in section 2.1.1 (see also Equation 2.4). The same behavior results independently of the coarse-graining level of the model [10, 11]. This universality is the reason why coarse-grained models can be employed.

Traditional polymer chain models

The wormlike chain is a quite realistic and intuitive model for semi-flexible chains: segments have a fixed contour length b and a finite bending stiffness κ. The persistence length over which the chain remains relatively straight is given by $\xi = b\kappa$.

The freely jointed chain is an ideal chain model, used often to study general properties of ideal chains on account of its simplicity. A chain consists of a sequence of N rigid segments, each of them having a length b, their orientation being completely uncorrelated. The end-to-end vector $\vec{R}_e$ is the sum of the individual segment vectors:

$$\vec{R}_e = \sum_{i=1}^{N} \vec{r}_i \qquad (2.1)$$

and for an ensemble of chains, we have the squared end-to-end distance

$$\langle \vec{R}_e^2 \rangle = \langle \left(\sum_{i=1}^{N} \vec{r}_i\right)^2 \rangle = \sum_{i=1}^{N} \langle \vec{r}_i^2 \rangle + 2 \sum_{i=1}^{N-1} \sum_{j=i+1}^{N} \langle \vec{r}_i \vec{r}_j \rangle \qquad (2.2)$$

Because $\langle \vec{r}_i^2 \rangle = b^2$,

$$\langle \vec{R}_e^2 \rangle = Nb^2 + 2b^2 \sum_{i=1}^{N-1} \sum_{j=i+1}^{N} \langle \cos\theta_{ij} \rangle \qquad (2.3)$$

where θ_{ij} is the angle between the vectors $\vec{r}_i$ and $\vec{r}_j$. θ_{ij} takes with equal probability any value between 0 and 2π. Therefore, $\langle \cos\theta_{ij} \rangle = 0$ and

$$\langle \vec{R}_e^2 \rangle = Nb^2. \qquad (2.4)$$

Lattice models: The monomer positions are confined to the sites of a lattice and the monomers are linked by bonds that connect, for example, nearest neighbor sites. This approach has some disadvantages due to the intrinsic anisotropy in such models, but it considerably simplifies both computer simulations and theoretical considerations.

The bead-spring model is an off-lattice chain model commonly used in computer simulations. The chains are represented by hard beads connected by elastic springs. A popular version of the model represents the beads by a shifted and truncated, purely repulsive Lennard-Jones potential connected by a so-called finitely extensible nonlinear elastic (FENE) potential [12, 13].

Soft-particle models

Certain physical phenomena occur on time and length scales larger than those related to monomers. Examples are the demixing of incompatible polymer blends or the microphase separation of block-copolymers. If, for example, we want to describe a process that takes place on the scale of the radius of gyration, degrees of freedom on the monomer scale will supposedly become irrelevant. To make simulations more efficient than they can possibly be by using coarse-grained molecular models like *e.g.* the previously mentioned bead-spring model, it is useful to employ models incorporating effective interactions on the scale of the radius of gyration and hence also capturing fluctuations on that scale.

It has been shown lately that the use of soft interactions of the type used in field theoretic descriptions results in very efficient simulations of homopolymers [14]. For the simulation of polymer chains in the melt, soft ellipsoid models have been employed [15, 16]. Very recently, it has been proposed to subdivide polymer chains into 10-100 subchains and represent them by soft spheres of fluctuating size related to the radius of gyration of the subchain [17].

There has been some research in recent years on soft dumbbell models (the dumbbell consisting of one soft sphere for each block) for block copolymer solutions that have been systematically coarse-grained by averaging over lattice model monomers [18, 19, 20] or by using probability distributions from Gaussian chains [21]. Based on liquid-state integral equations, a microscopic theory for coarse-graining diblock copolymers into such dumbbells has been developed recently [22].

2.2. Diblock copolymer melts

The homopolymer sequences of a block copolymer chain can be composed in a wide variety of architectures, e.g. linear copolymers, star copolymers, brush copolymers, and comb copolymers. The most basic system, consisting of one long sequence of the same unit bound to one long sequence of another unit, is the linear AB diblock copolymer. A schematic representation is given in Figure 2.1(a). In this work, we will focus on the structures formed by AB diblock copolymer melts. Due to its relative simplicity, the diblock copolymer melt system makes a good test candidate for both experiments and theoretic considerations.

Chemically distinct (incompatible) homopolymers tend to demix in a blend due to enthalpic incompatibility. If one links one end of a homopolymer to one end of a chemically distinct homopolymer, thus building a diblock copolymer, the covalent bond between the blocks prevents them from phase-separating on a macroscopic scale. However, they still tend to locally separate, but instead of separating macroscopically, they microphase separate and form periodic nanostructures. The microphase separation process is illustrated in Figure 2.1.

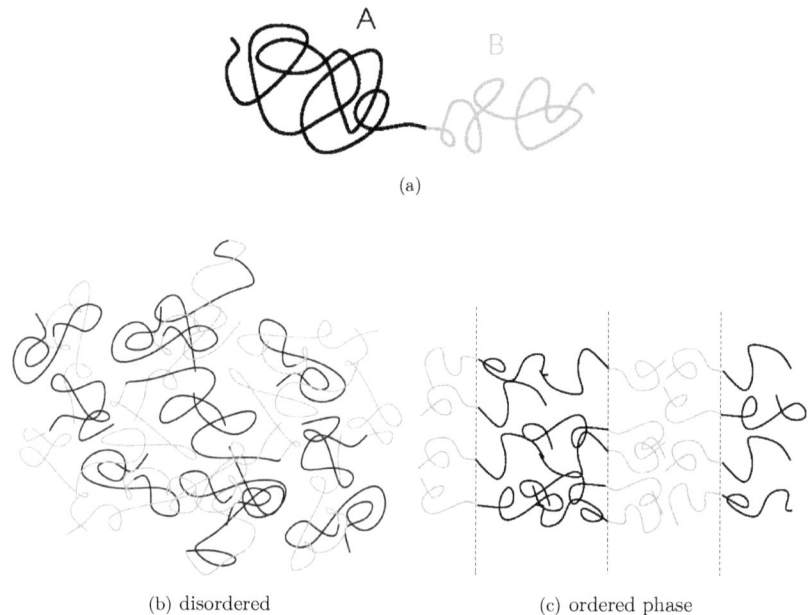

Figure 2.1.: (a) Architecture of a diblock copolymer chain. The microphase separation transition occurs when a compositionally disordered melt of copolymers (b) transforms to a spatially periodic, compositionally inhomogeneous phase (c) on lowering the temperature and thus increasing the repulsion between the blocks. For copolymer blocks with approximately the same length, the ordered phase has the lamellar structure that is shown in (c).

2.2 Diblock copolymer melts

2.2.1. Experimental methods

This section will give an overview of major methods that are commonly used in experiments to study the structural behavior of block copolymer melts experimentally [23].

Rheology (or dynamic mechanical spectroscopy) is employed to locate phase transitions. The temperature is increased slowly while the dynamic elastic shear modulus, G', is monitored at a fixed low frequency. Abrupt changes (*i.e.* a sharp decrease with rising temperature) in G' are associated with order-disorder phase transitions, because the flow properties of the melt depend on the state of order in the system: ordered phases are viscoelastic fluids whereas the disordered melt exhibits viscous flow [23].

SAXS (small angle X-ray scattering) and **SANS** (small angle neutron scattering) are ideal for the investigation of ordered phases of block copolymers, because they probe length scales that are typical also for block copolymer mesostructures (1 − 100nm). These techniques measure the scattering function (or structure factor) to determine the symmetry of the phase in question.

Whether X-rays or neutrons are used as incident radiation is dictated primarily by the contrast factor of the polymers that are used. Systems that exhibit a sizeable electron density difference between the components provide good X-ray contrast. Accordingly, these materials are frequently studied by SAXS.

The X-ray contrast is strongly reduced when structurally similar polymers are used, hence the use of SAXS is difficult or impossible. In that case, the neutron contrast can be increased by deuterating one of the blocks (*i.e.* exchanging ^{1}H by deuterium), making SANS the method of choice [24].

TEM (transmission electron microscopy) is the most direct method for the identification of block copolymer morphology. An image is formed from the interaction of the electrons that are transmitted through the sample. A disadvantage of this method is that misidentification of mesophases is possible because only a projection of a small sample is inspected [23].

AFM (atomic force microscopy) measures the deflection of a cantilever due to interactions between an attached sharp tip and the surface of a sample. While the sample is being scanned in x and y direction, this deflection is measured to provide an image of the surface structure. AFM can be used as a complementary tool to investigate materials which are too sensitive to be studied with an electron beam. When operated in dynamic mode, the cantilever is externally oscillated in z direction and the phase difference between the external excitation and the tip motion can be used to determine surface features and compositional variations [25, 26].

2.2.2. Theoretical concepts

Flory-Huggins theory for binary polymer blends

In 1942, Huggins [27] and Flory [28] independently introduced a simple mean-field lattice theory for polymer solutions and melts. It has since become the classical theory of phase separation in polymer blends and is fully derived in many textbooks [10, 29]. We briefly outline the derivation here to illustrate the origin of the Flory-Huggins parameter χ.

Polymer chains are represented as non-reversal random walks on a lattice. Consider a binary blend of homopolymers A and B with chain length (or *degree of polymerization*) N_A and N_B, and volume fractions φ_A and φ_B. If Ω is the total number of all lattice sites and n_A and n_B are the numbers of A and B chains, then the volume fractions are given by $\varphi_A = \frac{N_A n_A}{\Omega}$ and $\varphi_A = \frac{N_B n_B}{\Omega}$, $\varphi_A + \varphi_B = 1$.

To calculate the free energy $F = -k_B T \ln Z$, we need the partition function of the system:

$$Z = \sum_i \exp(-E_i/k_B T) \qquad (2.5)$$

where E_i is the energy of the configuration i of the polymers on the lattice. We estimate Z by replacing E_i by the average energy $\bar{E}$. The partition function then becomes

$$Z \approx W \exp(-\bar{E}/k_B T) \qquad (2.6)$$

where W is the total number of possible configurations of the polymers on the lattice.
$\bar{E}$ is determined by calculating the mean energy for the randomly mixed state, in which, on average, each lattice point is surrounded by $z\varphi_A$ A monomers and $z\varphi_B = z(\varphi_A - 1)$ B monomers. Hence the number of neighboring pairs of A monomers is given by $N_{AA} = n_A N_A z \varphi_A / 2 = z\Omega \varphi_A^2 / 2$, and in the same way, the number of neighboring pairs of B monomers is given by $N_{BB} = n_B N_B z \varphi_B / 2 = z\Omega \varphi_B^2 / 2$. Finally, the number of pairs of neighboring A and B monomers amounts to $N_{AB} = z\Omega \varphi_A \varphi_B$.

2.2 Diblock copolymer melts

The overall mean system energy for a given configuration which can be written as

$$\bar{E} = -N_{AA}\epsilon_{AA} - N_{BB}\epsilon_{BB} - N_{AB}\epsilon_{AB} \tag{2.7}$$

where the $\epsilon_{\alpha\beta}$ correspond to the interaction energies between neighboring monomers on the lattice, hence results in

$$\bar{E} \approx -z\Omega(\frac{1}{2}\varphi_A^2\epsilon_{AA} + \frac{1}{2}\varphi_B^2\epsilon_{BB} + \varphi_A\varphi_B\epsilon_{AB}). \tag{2.8}$$

With Equation (2.6), the free energy can then be obtained in the following form:

$$F = -k_B T \ln W + \bar{E}. \tag{2.9}$$

By considering the free energy of mixing, F_m, as the free energy of the mixed state minus the free energies of the pure components, it can be calculated to

$$F_m = \Omega k_B T \left(\frac{\varphi_A}{N_A} \ln(\varphi_A) + \frac{\varphi_B}{N_B} \ln(\varphi_B) + \chi\varphi_A\varphi_B \right). \tag{2.10}$$

The first two terms in the equation above represent the entropy of mixing of the two polymers, which is the same for the mixing of two ideal gases, except for the factor $1/N_{A,B}$, meaning that the entropic part of the equation (*i.e.* the part that favors mixing) is reduced by the chain length: the longer the molecules, the harder the mixing. The reason for this effect is that due to the linking of the segments the number of degrees of freedom decreases. The last part of Equation (2.10) accounts for the enthalpic incompatibility of the two components with the so-called Flory-Huggins parameter, χ, that is given by

$$\chi = \frac{z}{k_B T}\Delta\epsilon = \frac{z}{k_B T}(\epsilon_{AB} - \frac{1}{2}(\epsilon_{AA} + \epsilon_{BB})). \tag{2.11}$$

All microscopic chemical properties of the polymer mixture are expressed by this single parameter, which is therefore very hard to access. The "translation" of the χ parameter between theory, simulation and experiment, despite of being crucial for the interpretation and the quantitative relation of the phase diagrams, has not yet been well understood, and the establishing of an exhaustive theory is still a busy field of research with many unsolved questions [30, 31].

Self-consistent field theory

A more sophisticated mean-field approach is the self-consistent field theory (SCFT), first extended to block copolymers by Helfand and coworkers [32]. It evaluates the exact equilibrium behavior of a single test chain subjected to static mean fields that approximate the

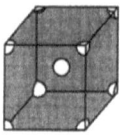 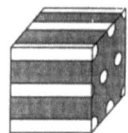   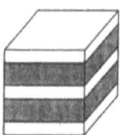

Figure 2.2.: Schematic representation of diblock copolymer mesophases: spheres (S), cylinders (C), gyroid (G), perforated lamellae (PL), lamellae (L) (from left to right).

total interaction with all the other chains. The values of these fields correspond to minima of the free energy, and fluctuations are ignored. These fluctuations are more important for short chains, and SCFT becomes exact in the $N \to \infty$ limit [11].

The self-consistent field equations can only be solved numerically, however approximate theories derived from them can have analytical solutions in certain limits, *cf.* section 2.2.3. A review on self-consistent field theory for block copolymer melts can be found in reference [33].

2.2.3. Phase diagram

When the incompatibility of unlike blocks becomes sufficiently high, the segments segregate into A- and B-rich domains forming periodically ordered microstructures like those shown in Figure 2.2. The ordering mechanism is illustrated for the lamellar mesophase in Figure 2.1. The resulting morphology depends on the composition of the diblock and the repulsion between the blocks that increases with decreasing temperature.

The simplest and most common structures include the lamellar (L) phase in which the domains form alternating flat layers, the cylindrical (C) phase where the minority component forms hexagonally arranged cylinders in a matrix made up of the majority component, and the spherical (S) phase where the minority blocks form spheres on a body-centered cubic (b.c.c.) lattice. There is also a wide variety of more complex microstructures, the most common being the gyroid (G) phase where the minority domain forms two interweaving three-fold coordinated lattices [34]. Experiments have also identified a perforated-lamellar (PL) phase resembling the lamellar phase, but with the minority lamellae perforated by hexagonally arranged holes through which the majority layers are connected.

2.2 Diblock copolymer melts

The first phase diagram for diblock copolymer melts was computed by Ludwik Leibler in 1980 [35]. He used an analytical approximation to the full mean field theory in the limit of weakly segregated melts, the so-called "random phase approximation" which is a perturbation calculation based on the completely disordered state.

This theory proved to be one of the most influential theories of the phase behavior of block copolymers. The state of the diblock copolymer system is determined by only two relevant parameters: the copolymer chain composition f, given by the fraction of the A block in the whole chain,

$$f = \frac{N_A}{N_A + N_B}$$

and the parameter that describes the interactions between both blocks, concerning which the product of the χ-parameter with the degree of polymerization, χN, is shown to be the relevant parameter rather than χ only.

The phase diagram calculated by Leibler predicts the lamellar, cylindrical ("hexagonal") and spherical ("b.c.c.") phases as most stable phases for the respective values of the parameters χN and f. Leibler found the order-disorder transition to be of first order at $f \neq 1/2$ and of second order at $f = 1/2$. However, the Leibler theory does not take into account fluctuations which have a strong influence especially at weak segregation [36], hence its predictive power has to be considered limited.

Matsen and coworkers [37] calculated a diblock copolymer phase diagram by full self-consistent field theory using the standard Gaussian model. This diagram is shown in Figure 2.3(b). It shows the morphology with the lowest free energy plotted as a function of segregation χN and composition f.

In addition to the mesophases that have already been discussed previously, an additional close-packed spheres phase (S_{cp}) is found.

We compare this SCFT phase diagram with a phase diagram constructed from experimental measurements of polystyrene-polyisoprene diblock copolymer melts which is shown in Figure 2.3(a). The dots in the experimental phase diagram correspond to the experimental data points.

To some extent, the same mesophases are found by experiments and mean field theories, and the agreement between both phase diagrams with respect to the range of morphologies and the sequence in which they occur is reasonable. The asymmetry in the experimental phase diagram is attributed to the chemical asymmetry between the segments of both

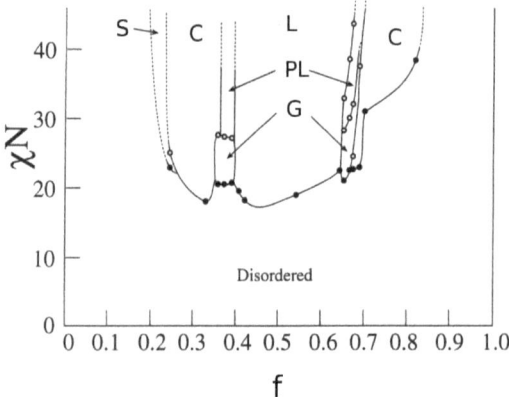

(a) Experimental phase diagram for polystyrene-polyisoprene diblock copolymer melts. Reprinted with permission from reference [38].

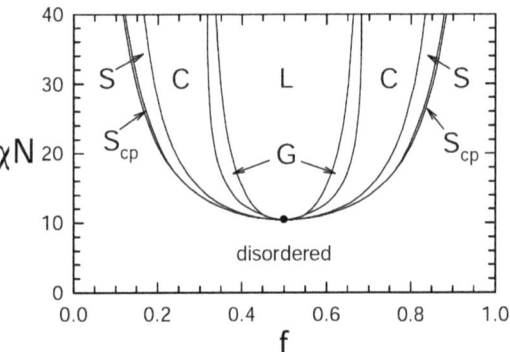

(b) SCFT phase diagram. Adapted with permission from reference [37].

Figure 2.3.: Comparison of the experimental phase diagram (a) with the phase diagram from self-consistent field theory (b).

2.2 Diblock copolymer melts

blocks. One striking difference is the absence of the perforated lamellar phase in the phase diagram predicted by SCFT. However, it has been indicated by careful experiments [39] that the PL state is a metastable state and eventually converts to the gyroid morphology. The actual morphologies of the phase diagrams differ, and there are significant qualitative deviations along the order-disorder transition. The experimental phase diagram shows direct first order transitions from the disordered into the ordered cylinder and gyroid phases, for example, whereas in the SCFT diagram, there is only a second order phase transition from the disordered into the lamellar phase at $f = 0.5$, and no direct disorder-order transitions into the cylinder or gyroid phase at all. The inaccuracies particularly close to the order-disorder transition are mainly due to the fact that mean-field theory neglects fluctuations, whereas the weakly segregated morphologies are most disrupted by fluctuations that are particularly strong near the critical point [36]. Therefore, fluctuations are also expected to increase the value of the χ parameter at which the order-disorder transition takes place. However one should remember that caution is always advised when comparing quantitatively values of the interaction parameter χ from SCFT with experiments or simulations, because there is no properly defined analogy, as has been stated previously in this chapter.

One way to account for fluctuation effects that has been pursued in the last decades is the extension of mean field theories to incorporate fluctuations [41, 42] or to include conformational fluctuations of the chains in a hybrid particle-based "single chain in mean field" (SCMF) approach [43, 44]. Another approach is the use of molecular computer simulations.

A partial phase diagram (shown in Figure 2.4) was constructed by Matsen et al. using Monte Carlo simulations of an fcc-lattice model [40]. This study was limited to the location of the order-disorder transition, but the results were summarized in the form of a phase diagram for convenient comparison with those in Figure 2.3.
In this study, Matsen et al. used cooling and heating runs to address several values of incompatibility. This produced a hysteresis loop that enabled them to locate the order-disorder transition in the interval defined by the loop. Similar to the experimental phase diagram, direct phase transitions into the ordered phases were found for all f due to the fact that fluctuations are incorporated.

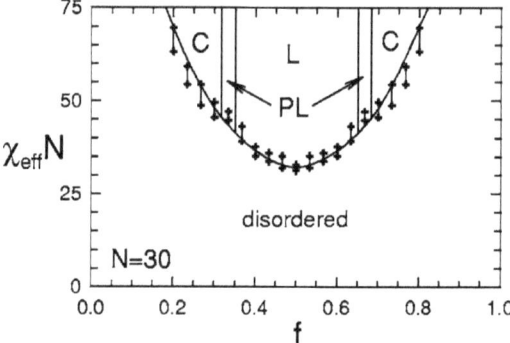

Figure 2.4.: Partial phase diagram using lattice-based Monte-Carlo simulations. The effective interaction parameter χ_{eff} is proportional to the lattice occupancy, no attempt is made to account for the difference between the lattice and continuum definitions of χ which is necessary for a quantitative comparison with Figure 2.3. Reproduced with permission from reference [40].

CHAPTER 3

Model

In this work, we introduce a new soft-particle-type non-lattice model for computer simulations of diblock copolymer melts. The details of the model will be described in this chapter.

3.1. Introduction

Traditional coarse-grained models like *e.g.* the bead spring model are still too detailed for the simulation of certain phenomena that occur on larger scales. The microphase separation in block copolymers happens on a length scale which is comparable to the radius of gyration of the blocks, hence it should be possible to simulate this phenomenon with a model working with effective interactions on that same length scale. This allows computationally more efficient simulations, compared to the above mentioned traditional models.
To justify this approach, we use the fact that in a melt, the behavior of a polymer chain can be described as that of an ideal (*i.e.* Gaussian) chain, as has been motivated in chapter 2.1.1.
In a melt in three spatial dimensions, polymer chains significantly interpenetrate. Within the radius of gyration of a chain, however, the probability of finding monomers of other chains is smaller than the average probability. This "correlation hole effect" can be trans-

lated into an effective soft repulsive potential with a range of about $2R_g$ and an amplitude of about $2k_BT$, as references [45, 46] found out for dilute and semi-dilute solutions. The amplitude stays approximately the same for all investigated concentrations, extending deep into the semi-dilute regime at concentrations of the order of ten times the overlap concentration. The centers of mass of polymer chains have a gas-like pair distribution function for all concentrations, from the dilute regime extending up to the melt regime. Thus, we expect the general behavior of an effective potential that has been derived for dilute and semi-dilute solutions to be also valid for the melt.

The tensor of inertia of a single chain in the melt has three principle moments E_i, with $E_1 > E_2 > E_3$. This corresponds to an ellipsoidal envelope of the instantaneous chain conformation [47, 15].

In order to mimic this non-spherical instantaneous shape, but still have the advantage of the simpler treatment of spheres, we map each of the two blocks onto a dumbbell, *i.e.* two spheres. Thereby we assume that half of each block in the melt can still be described by a Gaussian chain.

Since a diblock chain is thus described by four spheres, we call it a *tetramer*. Because polymer chains in melts overlap significantly, the interaction between these spheres has to be very soft to permit them to interpenetrate.

References [18, 19, 20] derive effective interactions between centers of "soft blobs" by averaging over monomer degrees of freedom and state that the resulting effective potentials have "roughly Gaussian" form [20]. Although for analytical treatment, a Gaussian might be more suitable, for numerical treatment, it is preferable to have an interaction of strictly finite range to avoid cut-off effects.

To capture qualitatively the essential features of the potentials systematically derived in the references that have been quoted above, our potential has to be parabolic near the origin and go to zero around $2R_g$ [45, 46]. As a non-bonded interaction potential, we choose a third order polynomial that is cut off at a length σ given by the first root of the polynomial. This choice mimics the bell curve shape of the potentials that have been derived in references [18, 19, 20] but has the advantage that it goes exactly to zero at a well-defined distance which therefore provides a natural cut-off length.

The effective potentials that are used in models on this scale are potentials of mean force rather than conventional pair potentials, and they are therefore generally temperature dependent. According to references [45, 46], the effective potential acting between soft spheres of the same species should have an amplitude of about $2k_B T$. In order to allow the system to exhibit microphase separation, a parameter is needed that controls the repulsion between different species. One can either keep the temperature constant and vary the parameter that describes the repulsion strength between A and B, or, which is equivalent, one can make the amplitude of the repulsion between different components independent of temperature, while we choose the amplitude of the effective repulsive potential acting between spheres of the same species to decline with decreasing temperature. This means that, to simplify matters, we neglect entropic contributions for the interaction between spheres of unlike species, and make this interaction purely enthalpic.

3.2. Model Description

In this section, we describe the properties of the soft-tetramer model for computer simulations of molten diblock copolymer chains consisting of two species A and B.

3.2.1. Interaction potentials

The potential for non-bonded interactions is purely repulsive and acts over a range of σ_{AA}, σ_{BB}, or $\sigma_{AB} = \frac{1}{2}(\sigma_{AA} + \sigma_{BB})$, respectively. The interaction between spheres of the same species is entropic, and we set its amplitude $\epsilon_{\alpha\alpha} = 2T$ (cf. chapter 3.1, for simplicity's sake we set $k_B = 1$):

$$U^{\text{nb}}_{\alpha\alpha}(r) = \begin{cases} \epsilon_{\alpha\alpha}\left(1 - 3\left(\frac{r}{\sigma_{\alpha\alpha}}\right)^2 + 2\left(\frac{r}{\sigma_{\alpha\alpha}}\right)^3\right) & r < \sigma_{\alpha\alpha}, \\ 0 & r > \sigma_{\alpha\alpha}. \end{cases} \quad (3.1)$$

where $\alpha \in \{A, B\}$ and r is the distance between the centers of the spheres.

Interaction between spheres of unlike species have enthalpic character, so we choose the pair potential for unlike species to have the same form but to be independent of T:

$$U^{\text{nb}}_{AB}(r) = \begin{cases} \epsilon_{AB}\left(1 - 3\left(\frac{r}{\sigma_{AB}}\right)^2 + 2\left(\frac{r}{\sigma_{AB}}\right)^3\right) & r < \sigma_{AB}, \\ 0 & r > \sigma_{AB}, \end{cases} \quad (3.2)$$

with $\epsilon_{AB} = 2$. With an interaction potential chosen like this, there will be effective repulsion between unlike species with decreasing temperature, enabling the system to exhibit phase separation. At $T = 1$, interactions between like and unlike species are equal.

For bonded interactions, a harmonic term is added to the non-bonded interaction potential:

$$U_{\alpha\alpha}^{\text{bond}}(r) = U_{\alpha\alpha}^{\text{nb}}(r) + 1.6\, T \left(\frac{r}{\sigma_{\alpha\alpha}}\right)^2 \qquad (3.3)$$

$$U_{AB}^{\text{bond}}(r) = U_{AB}^{\text{nb}}(r) + 1.6 \left(\frac{r}{\sigma_{AB}}\right)^2 \qquad (3.4)$$

This choice of bond potential leads to a purely enthalpic bonded AB interaction. Reference [21] shows that the probability distribution for the distance between two spheres representing the A and B block of a diblock copolymer molecule forms a three-dimensional Gaussian with an average distance in the order of the size of the spheres. Most notably, it does not depend significantly on temperature. In our case, the size of the spheres is fixed, so we can assume the average separation between the spheres as constant and we reproduce the behavior found in reference [21] by making the AB interaction completely temperature independent.

This description is equivalent to a description where the temperature is kept constant to $T = 1$ and the parameter $\tilde{\epsilon}_{AB}$, given by $\epsilon_{AB} = \tilde{\epsilon}_{AB} T$, is varied.

Figure 3.1 illustrates the idea of the soft-tetramer model. The parameters σ_{AA} and σ_{BB} can be interpreted as the diameters of the respective spheres making up the diblock chain. They are designed to correspond to about $2R_g$ of the underlying polymer chains that are represented by this model.

3.2.2. Control parameters

The two relevant control parameters for the mean-field diblock phase diagram are the composition of the chain, $f = N_A/(N_A + N_B)$, and χN, the product of the Flory-Huggins χ-parameter and the chain length N, cf. chapter 2.2.2.

In the soft-tetramer model, differences in the chain composition f are translated into a different ratio of radii of A and B spheres. Since we assume Gaussian statistics for each block, and thus $\sigma_{\alpha\alpha}^2 \propto b^2 N_\alpha$ with the same statistical segment length $b_A = b_B = b$ for both

3.2 Model Description

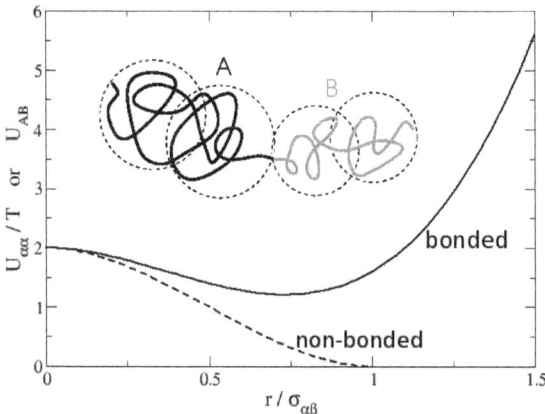

Figure 3.1.: Bonded and non-bonded interaction potentials. The inset shows a schematic representation of the model (not to scale).

blocks, f can be determined as follows:

$$f = \frac{N_A}{N_A + N_B} = \frac{b^2 N_A}{b^2 N_A + b^2 N_B} = \frac{\sigma_{AA}^2}{\sigma_{AA}^2 + \sigma_{BB}^2} \tag{3.5}$$

We keep the chain length N constant, which translates to

$$\sigma_{AA}^2 + \sigma_{BB}^2 = 2\sigma^2, \tag{3.6}$$

where σ corresponds to the diameter of the spheres in the compositionally symmetric case, $f = 0.5$. With

$$\sigma = 1 \tag{3.7}$$

as a natural choice of length unit in our model, we obtain

$$\sigma_{AA} = \sqrt{2f} \tag{3.8}$$

$$\sigma_{BB} = \sqrt{2 - 2f} \tag{3.9}$$

For symmetry reasons, we only consider cases with $0.5 < f < 1$, making A the majority component.

The Flory-Huggins χ-parameter, originating from a mean-field theory built on simplifying approximations, has no properly defined analogon in non-mean-field models [48, 49] or in experiment [50] (refer also to the related discussion in chapter 2.2.2). The phase separation process is driven by the chosen pair interactions, or, more precisely, by their asymmetry

$$\Delta\epsilon = \epsilon_{AB} - \frac{1}{2}(\epsilon_{AA} + \epsilon_{BB}), \qquad (3.10)$$

that characterizes the repulsion between A and B segments. It is generally assumed that $\chi \propto \Delta\epsilon$ with the proportionality depending on z which in lattice simulation is given by the coordination number[1] of the lattice type and in continuum simulations by the average number of interchain neighbors. However, this is a very simplifying assumption, and doubts about its validity have been raised from simulations [48].

Thus, without making an attempt to relate our interaction asymmetry parameter χN to the mean field χ parameter, we define it in our model in the following way:

$$\chi N = 4\frac{\Delta\epsilon}{T} = \frac{4(2-2T)}{T} = 8\left(\frac{1}{T}-1\right). \qquad (3.11)$$

Here, N is taken to be the number of soft spheres, 4.
Note that $T = 1$ in our model corresponds to the completely symmetric case: $\chi = 0$.

3.3. Invariant degree of polymerization

A useful parameter to compare experiments with mean-field calculations and simulations is the invariant degree of polymerization N_{inv}, because it is independent of the chain discretization, i.e. the definition of a monomer unit. It is a measure for the number of neighbors with which a polymer interacts. The overlap of molecules averages out fluctuation effects in the $N_{\text{inv}} \to \infty$ limit, meaning that mean-field theory would become exact in that limit.
The invariant degree of polymerization is defined as

$$N_{\text{inv}} = (\rho_c R_e^3)^2 = \left(\rho_0 \frac{R_e^3}{N}\right)^2 \qquad (3.12)$$

---

[1] In some studies, the coordination number is reduced by two to account for the fact that the interactions between neighboring monomers along the chain are fixed. However, this is of disadvantage for small N_A, N_B. This introduces a further ambiguity in the interpretation of the χ parameter.

3.3 Invariant degree of polymerization

where ρ_c is the chain number density, ρ_0 is the monomer number density and R_e is the distance between the chain ends, the so-called end-to-end distance. For a Gaussian chain,

$$R_e^2 = b^2 N \tag{3.13}$$

as has been shown in Equation (2.4).

We now write the length of the whole chain as the sum of the chain lengths of both blocks,

$$N = N_A + N_B \tag{3.14}$$

and remember that in our model the following relation holds true, as we have motivated previously:

$$\sigma_{\alpha\alpha} = 2R_{g\alpha} \tag{3.15}$$

with $\alpha \in \{A, B\}$. Furthermore, for the Gaussian chain, the following relation applies:

$$R_g^2 = \frac{1}{6} N b^2 \tag{3.16}$$

Hence, combining Equations (3.15) and (3.16) while considering the fact that one soft sphere of type α represents $\frac{N_\alpha}{2}$ segments, we obtain

$$\sigma_{\alpha\alpha}^2 = 4R_{g\alpha}^2 = \frac{1}{3} N_\alpha b^2 \tag{3.17}$$

and thus, using Equation (3.6), we can conclude that in our model, using the corresponding length scale defined in Equation (3.7), we have:

$$R_e^2 = b^2(N_A + N_B) = 3(\sigma_{AA}^2 + \sigma_{BB}^2) = 6. \tag{3.18}$$

Finally, if we use the relation above in Equation (3.12), we can determine the invariant degree of polymerization in the soft-tetramer model:

$$N_{\text{inv}} = (R_e^2)^3 \rho_c^2 = 216 \rho_c^2 \tag{3.19}$$

Values reached in experiments typically are in the order of $N_{\text{inv}} = O(10^4)$. These values can also be reached in SCMF calculations, realized by increasing the polymer density ρ_c, i.e. the number of chains in the ensemble [44]. Molecular simulations, however, usually reach values for N_{inv} which are smaller by two orders of magnitude. This is also the case for our simulations.

CHAPTER 4

Geometric characterization of mesophases

A method that is used commonly in experiments [38, 23] and simulations [40] to determine the structure of the diblock copolymer mesophase at a certain set of control parameters is what one could call *direct geometric observation*. In experiments, this is done by looking at *e.g.* TEM or AFM images (*cf.* also chapter 2.2.1), in simulations by visualizing configurations in a simulation snapshot. When we consider a simulation snapshot like the one shown in Figure 4.1, we can clearly identify the present phase, in that case as the lamellar state.

This chapter will describe techniques to formalize and quantify this geometric approach to mesophase identification and discuss the observables that are employed for the determination of the morphology of the simulated diblock copolymer melt.

4.1. Mesophase identification by cluster analysis

The basic idea of the method that will be introduced in this chapter is the objective characterization of block copolymer mesophases by quantifying the approach of direct geometric observation.
What we basically do when we identify structures by the eye is that we look separately at clusters of one component and then decide which shape these clusters have. The method

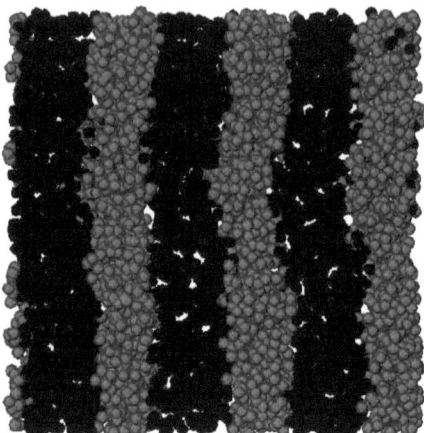

Figure 4.1.: Example simulation snapshot, lamellar mesophase.

we want to apply follows the same idea: First we identify clusters of neighboring spheres of the same component, and then we determine the shape of these cluster by calculating the eigenvalues of their gyration tensors.

4.1.1. Cluster identification algorithm

Clusters of neighboring tetramer spheres belonging to the same species respectively are identified via a Hoshen-Kopelman based algorithm for non-lattice environments. Using specific properties of these clusters like the numbers of majority and minority clusters and their gyration tensors, all observed phases can be identified.

4.1.1.1. Hoshen Kopelman algorithm and extension to non-lattice environments

The Hoshen-Kopelman algorithm, originally described by J. Hoshen and R. Kopelman in 1976 [51], is an efficient technique for identifying clusters of contiguous network nodes. The original algorithm identifies clusters of occupied sites on a regular lattice. It has been

4.1 Mesophase identification by cluster analysis

extended by Al-Futaisi and Patzek for a non-lattice environment where the sites can be placed randomly at non-lattice points and each site can have arbitrary connectivity [52]. We adapted this algorithm for the specific conditions of the soft-tetramer model. The network nodes or sites are defined by the centers of the soft spheres. We call two nodes "connected" when they have a separation of less than a given distance r_{nn}. The choice of this distance is justified in section 4.1.1.2. The first step in this continuum version of the algorithm is therefore the preparation of a list of all nodes (separately for the components A and B) and a list of same-type neighbors for every node. Note that for the cluster determination, we do not apply the periodic boundary conditions that are used in the simulation. Only the configuration within the simulation box is considered.

Then, the clusters are determined (again, separately for A and B), which means that the correct cluster number has to be assigned to each node. This is done in the following way:

1. A cluster counter variable n_c is set to one.

2. The network is traversed node by node, using the list of all nodes, and for each node, first we check if a cluster number has already been assigned. If yes, we continue with the next node in the list of nodes (go to step 2.) If not, we continue with step 3.

3. We set the cluster label of this node to the current value of the cluster counter variable n_c. Then we traverse the node's list of neighbors node by node, and check for each neighbor node whether it has already been assigned a cluster label. If this is the case, we go to the next node in the neighbor list. If it has not been assigned a cluster label yet, we set the label of this node equal to the current value of the cluster counter variable and traverse its neighbor list node by node, and so on, recursively.

4. At some point, the recursive algorithm reaches a "dead end", where there are no unlabeled neighboring nodes left. In that case, we continue with passing through the next higher level of node list.

5. When the algorithm reaches the first level again, $i.e.$ the list of all nodes, this means that the n_cth cluster has been completely labeled. n_c is increased by one and we continue with step 2.

When the algorithm has stopped in the end after scanning the whole network, we have identified $(n_c - 1)$ clusters.

Geometric characterization of mesophases

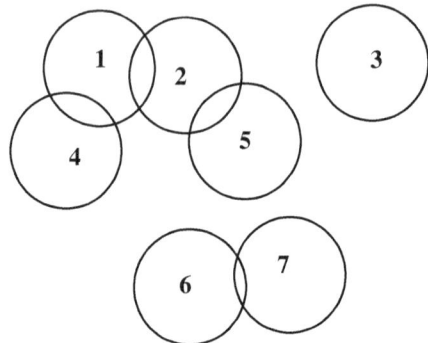

Figure 4.2.: Network example consisting of seven nodes of the same species.

We will illustrate the algorithm by a simple example: Consider the network shown in Figure 4.2. Connected nodes are shown as intersecting circles, so we clearly can identify three clusters here: The largest one consisting of nodes 1, 2, 4 and 5, the second one consisting of nodes 6 and 7 and last, a single-node cluster consisting only of node 3. Let us step through the algorithm for this example network:

1. Set $n_c = 1$. We traverse the list of nodes in numerical order.

2. Since node 1 is not labeled yet, it is now assigned the cluster number n_c, *i.e.* one.

3. The list of neighbors of node 1, containing node 2 and 4, is scanned.

4. Node 2 is not labeled yet, so its label is now set to one. Its neighbor list, which contains nodes 1 and 5, is traversed.

5. Node 1 is already labeled, so we go directly to the next node in the list, which is node 5. Node 5 is not labeled yet, so its cluster label is set to one.

6. Node 5 only has node 2 in its neighbor list, which is already labeled, and since we already finished the scanning of node 2's neighbor list, we continue with scanning the neighbor list of node 1.

7. The only node left in node 1's neighbor list is the yet unlabeled node 4, so it is assigned cluster label one.

4.1 Mesophase identification by cluster analysis

8. Node 4 has only node 1 as a neighbor, which already has a cluster label, hence the algorithm has created the first cluster and n_c is increased by one.

9. We continue traversing the list of all nodes. As node 2 has already been assigned a cluster label, we continue with node 3 and assign to it the cluster number equal to the current n_c, i.e. two. Since node 3's neighbor list is empty, we continue traversing the list of nodes after increasing the cluster counter n_c by one.

10. For nodes 4 and 5, no action is taken because they have already been assigned to existing clusters.

11. Node 6 has not got a cluster label yet, so it is assigned the current value of n_c, i.e. three. The neighbor list contains only node 7, which is also not labeled yet, so it also gets cluster number three. Node 7's neighbor list only contains the already labeled node 6, so we return to the list-of-nodes level, and the cluster counter is increased to $n_c = 4$.

Since there are no nodes left in the list of nodes, the algorithm stops here, after detecting $n_c - 1 = 3$ clusters.

4.1.1.2. Choice of next-neighbor radius

In order to explain the choice of next-neighbor radii, we have to anticipate simulation results that will be discussed later in more detail.

To determine the optimal choice of the distance at which two spheres still belong to the same cluster, we compared four different sets of next neighbor radii:

1. $r_{nn,A} = \sigma_{AA}$, $r_{nn,B} = \sigma_{BB}$,

2. $r_{nn,A} = r_{nn,B} = \sigma_{AA}$,

3. $r_{nn,A} = r_{nn,B} = \sigma_{BB}$,

4. $r_{nn,A} = r_{nn,B} = \sigma_{AB}$.

By comparing the distributions of the numbers of majority and minority clusters that we obtain for the parameters $f = 0.78$, $\chi N = 42$ (corresponding to the cylindrical phase as determined in the traditional way by looking at the corresponding simulation snapshot in Figure 4.3(a)) which have been calculated with the different choices of next neighbor radii, respectively (see Figure 4.4), we observe that cases (1.) and (2.) and in first approximation also (4.) give very similar results, whereas case (3.) is completely off from what we expect for the cylindrical phase (as described in section 4.1.2.4), because the expected single majority cluster decomposes in even more clusters than the minority component. If one looks closely at case (4.), one can see that here this effect is in fact much weaker, but the majority cluster already starts to decompose in up to four sub-clusters.

We can thus focus on the first two cases for the selection of the optimal next neighbor radii. For this purpose, we compare the two cases again for two different parameters: $f = 0.75$, $\chi N = 21$. This corresponds to a perforated lamellar phase as can be seen from the simulation snapshot in Figure 4.3(c)). The results are shown in Figure 4.5. Here, we can see that choice (1.) better separates the minority lamellae. Concluding the results from this comparison we choose the next neighbor radii for the cluster analysis algorithm to be as given in case (1.).

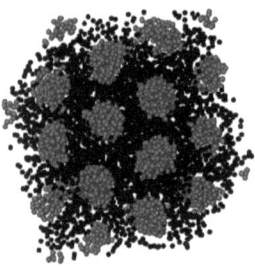
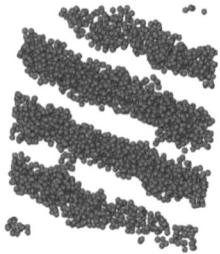
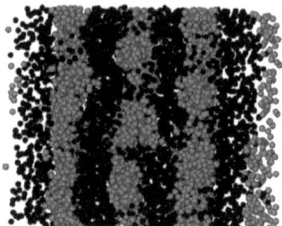

(a) Head-on view onto the plane perpendicular to the cylinders

(b) Side view, only B blocks are shown

(c) Perforated lamellae phase

Figure 4.3.: Simulation snapshot of the cylinder phase at $\rho_c = 1.3$, $f = 0.78$ and $\chi N = 42$ ((a) and (b)) and of the perforated lamella phase at $\rho_c = 1.3$, $f = 0.72$ and $\chi N = 21$ (c).

4.1.2. Mesophase identification

After identifying the clusters in a given configuration sample, we use specific cluster properties to determine the phase, e.g. the shape and the number of the clusters. Here we

4.1 Mesophase identification by cluster analysis

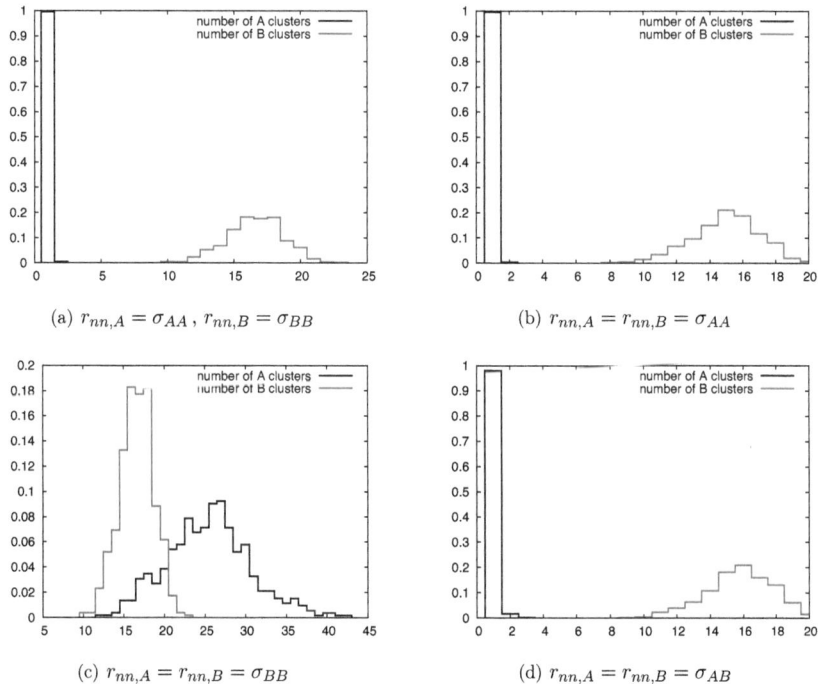

Figure 4.4.: Comparison for choices (1.)–(4.) of next neighbour radii in the cylindrical phase, parameters $f = 0.78, \chi N = 42, \rho_c = 1.3$.

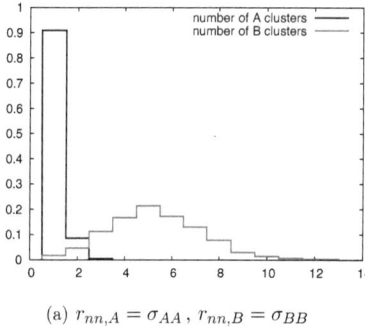
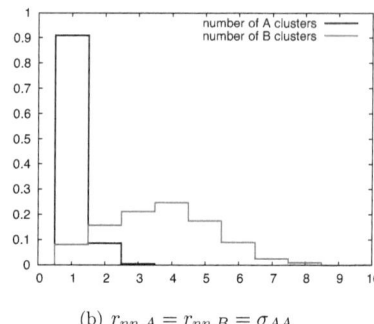

(a) $r_{nn,A} = \sigma_{AA}$, $r_{nn,B} = \sigma_{BB}$ (b) $r_{nn,A} = r_{nn,B} = \sigma_{AA}$

Figure 4.5.: Recomparison for choices (1.)–(2.) of next neighbor radii in the perforated lamellar phase, parameters $f = 0.75, \chi N = 21, \rho_c = 1.3$.

only take into account clusters that consist of more than three spheres. Note that in the following, when we use the term "cluster", we will always be referring to clusters with more than three spheres.

The shape is determined by calculating the gyration tensor of each cluster

$$G_{\mu\nu} = \frac{1}{N_p} \sum_{i=1}^{N_p} r^i_\mu r^i_\nu, \quad \mu, \nu \in \{x, y, z\}, \qquad (4.1)$$

where the r_μ are the coordinates of the particle in the center of mass reference frame of the cluster, and the sum runs over all particles which constitute the cluster.

After this calculation, the tensor is diagonalized to determine its eigenvalues G_μ. The eigenvalues are then ordered so that $G_1 > G_2 > G_3$.

The numbers of majority and minority clusters in combination with the eigenvalues of the gyration tensor characterize the phase in a distinctive way. In the following, we will explain how this characterization works.

4.1.2.1. Lamellar phase

In the lamellar state, we expect to find about the same numbers of minority and majority clusters. The clusters ideally should take the form of cuboids, so we compare the eigenvalues of the gyration tensor of the clusters to those of an ideal cuboid which we calculate with

4.1 Mesophase identification by cluster analysis

the "continuum version" of Equation (4.1):

$$G_{\mu\nu} = \frac{1}{V} \int dV\, r_\mu r_\nu, \tag{4.2}$$

For the gyration tensor of an ideal cuboid with side lengths L_1, L_2 and L_3, we obtain:

$$(G_{cub}) = \frac{1}{12} \begin{pmatrix} L_1^2 & 0 & 0 \\ 0 & L_2^2 & 0 \\ 0 & 0 & L_3^2 \end{pmatrix}. \tag{4.3}$$

4.1.2.2. Perforated lamellar phase

The perforated lamellar phase has only one big majority cluster because the majority layers are connected to each other through the holes in the perforated minority clusters, while the distribution of the number of minority clusters is similar to that in the lamellar phase. Accordingly, the gyration tensor of the majority phase is given by the size and shape of the simulation box:

$$(G_{cub}) = \frac{1}{12} \begin{pmatrix} L_x^2 & 0 & 0 \\ 0 & L_y^2 & 0 \\ 0 & 0 & L_z^2 \end{pmatrix} \tag{4.4}$$

where L_x, L_y and L_z are the linear dimensions of the simulation box in direction x, y and z respectively. In our bulk simulations, the simulation box is always cubic, so $G_1 = G_2 = G_3$. The gyration tensor of the minority phase resembles that of the lamellar phase given in Equation (4.3).

4.1.2.3. Gyroid phase

For the gyroid phase we should ideally find one majority and two minority clusters since in that case, the minority component forms two distinct channels that never intersect [53]. All eigenvalues of the gyration tensors of all three clusters are dictated by the size and the shape of the simulation box, similar to the case of the majority cluster in the perforated lamellar phase. Hence, the gyration tensors of all three clusters ideally look like the one given in Equation (4.4).

4.1.2.4. Cylindrical phase

In the cylindrical phase, we expect to find one big majority cluster and several small minority clusters. The majority cluster approximately takes the shape and the size of the box (again, with the gyration tensor dictated by the box dimensions, see Equation (4.4)) whereas the form of the minority clusters can be compared to cylinders, with the gyration tensor ideally given by:

$$(G_{cyl}) = \begin{pmatrix} \frac{1}{12}L^2 & 0 & 0 \\ 0 & \frac{1}{4}R^2 & 0 \\ 0 & 0 & \frac{1}{4}R^2 \end{pmatrix} \quad (4.5)$$

where R is the radius and L the length of the cylinder.

4.1.2.5. Spherical phase

In the spherical phase, we will again find one majority cluster extending over the complete simulation box and in the order of 50 minority clusters (for our simulation box sizes). The gyration tensor of the spherical minority clusters has three equal principle axes and is ideally given by:

$$(G_{sphere}) = \frac{2}{5} \begin{pmatrix} R^2 & 0 & 0 \\ 0 & R^2 & 0 \\ 0 & 0 & R^2 \end{pmatrix} \quad (4.6)$$

where R is now the radius of the spherical structure formed by the minority component.

4.1.3. Shape parameters

We will also analyze two shape parameters K_1 and K_2 that can be calculated from the eigenvalues of the gyration tensor and which together give information about the form of the cluster [54]. They are defined as

$$K_1 = \frac{G_2 + G_3}{G_1 + G_2} \quad (4.7)$$
$$K_2 = \frac{G_1 + G_3}{G_1 + G_2}.$$

The point $(K_1 = 0, K_2 = 1)$ is the ideal rod, $(K_1 = 1/2, K_2 = 1/2)$ is the ideal disk and $(K_1 = 1, K_2 = 1)$ is the ideal sphere.

4.2 Minkowski functionals

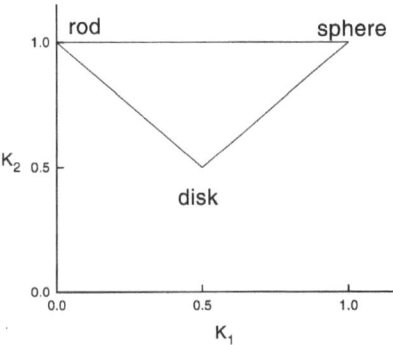

Figure 4.6.: Location of ideal solids in the (K_1, K_2)-plane.

A transition from a disk-like lamellar shape of the minority component clusters to a rod-like cylinder shape, for example, should show a movement of the corresponding point in the (K_1, K_2)-plane from the region around $(1/2, 1/2)$ towards $(0, 1)$.

4.2. Minkowski functionals

Minkowski functionals [55] are additive image functionals which can be employed to characterize a given pattern by objective numerical values. The calculation of Minkowski functionals is therefore a widely used method in morphological image analysis [56], hence we also investigated if this could be a useful approach to the determination of our soft sphere model block copolymer mesophases.

4.2.1. Integral geometry

Integral geometry is concerned with measures M for spatial distributions A which fulfill the following conditions:

- additivity, meaning $M(A_1 \cup A_2) = M(A_1) + M(A_2) - M(A_1 \cap A_2)$.
- invariance with respect to rotation

- invariance with respect to translation.

Minkowski functionals constitute a complete basis for all measures that fulfill these conditions. In d-dimensional space there are exactly $d+1$ of such functionals, hence for $d=3$, there are four Minkowski functionals: the volume V, the surface area S, the mean breadth B and the Euler characteristic χ_e. Usually, the Euler characteristic is referred to as χ, but because this letter is already occupied in the frame of this work by the incompatibility parameter (see definition in section 3.2.2), we will refer to it in this work as χ_e.

As opposed to many works in literature, we calculate the Euler characteristic of a body A, and not of the surface ∂A. However, both are closely related, namely

$$\chi_e(\partial A) = [1 - (-1)^n]\, \chi_e(A), \tag{4.8}$$

where n is the dimension of A [56].

4.2.2. Mapping and calculation

For a three-dimensional lattice filled with black and white cubic voxels, Minkowski functionals can be determined easily by simple counting of vertices, edges, faces and bodies of the voxels.

In order to calculate Minkowski functionals for our configurations, as a first step we thus have to subdivide the whole simulation box into cubic voxels and map the structures to the resulting three-dimensional lattice, *i.e.* transform the structure into a discrete three-dimensional black-and-white picture, or more exactly an *"A-and-B"* picture.

We calculate the following quantity in order to decide if voxel j is to be colored "A" or "B":

$$V_j = \sum_i^{N_S} d(\text{type}) \cdot U_{AB}^{nb}(|\vec{r}_{ij}|), \qquad d(\text{type}) = \begin{cases} +1, & \text{if type} = A, \\ -1, & \text{if type} = B \end{cases} \tag{4.9}$$

where $|\vec{r}_{ij}|$ is the distance between sphere i and the center of volume element j, and $U_{AB}^{nb}(|\vec{r}_{ij}|)$ is the value of the non-bonded potential that would act between sphere i and a sphere of unlike type located at the center of volume element j (see Equation (3.2)). The sum runs over the total number of spheres in the box N_s. If $V_j > 0$, the volume element is considered an A voxel, if $V_j < 0$, it is considered a B voxel.

Figure 4.7 demonstrates the mapping procedure for a simple two-dimensional example case with only two spheres in the relevant area.

4.2 Minkowski functionals

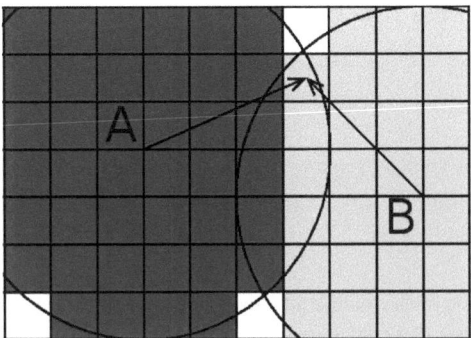

Figure 4.7.: Mapping schematics for a simple two-dimensional two-sphere example. The arrows indicate the vector $\vec{r}_{ij}$.

In the spirit of our approach that mimics the process of direct geometric observation, we do not apply periodic boundary conditions when we calculate Minkowski functionals. The three-dimensional images that we investigate consist of the configurations inside of the simulation box.

The procedure to calculate Minkowski functionals consists of decomposing each of the pattern constituting cubic voxels into 8 vertices, 12 edges, 6 faces and 1 cube, and then counting their total numbers, where the interfaces of adjoining voxels may only be counted once. Afterwards, the Minkowski functionals can be calculated as follows:

$$M_1 = V = n_c, \tag{4.10}$$

$$M_2 = S = -6n_c + 2n_f, \tag{4.11}$$

$$M_3 = 2B = 3n_c - 2n_f + n_e, \tag{4.12}$$

$$M_4 = \chi_e = -n_c + n_f - n_e + n_v \tag{4.13}$$

where n_c, n_f, n_e and n_v are the total numbers of cubes, faces, edges and vertices, respectively.

4.2.3. Minkowski functionals for diblock copolymer mesophases

In the following, we will discuss the usefulness of Minkowski functionals for the determination of block copolymer mesophases.

The volume V is trivially given by the respective volume fraction of the component we are investigating, hence it is basically constant at a given composition f and can not be used for phase identification.

The discretization of the images introduces cubic distortions in the objects, which result in a systematic error that afflicts the surface area S calculated from the distorted object, even in the limit of infinitely high resolution.
This effect is shown by the following experiment: We map a cubic box with length $L = 12$ containing four ideal black cylinders with radius $r = 1.5$ and length $l = 12$ in a white matrix to $n \times n \times n$ cubic cells, increase n gradually and calculate the surface for each n using Equation (4.11).
The surface vs. linear discretization n is shown in Figure 4.8. While the ideal value of the surface of four cylinders with radius $r = 1.5$ and length $l = 12$ should be $S = 4\left(2\pi rl + r^2\pi\right) \approx 509$, we see that the value calculated from the mapped image converges to $S \approx 635$ for high values of n.
Since even for ideal cylinders, the method yields an incorrect result, the surface area, too, is not expected to be useful for the identification of the mesophases, especially in the presence of fluctuations resulting in a broad distribution of geometric characteristics of the structures.

To gain a physical understanding of the the third and fourth Minkowski functional, the mean breadth B and the Euler characteristic χ_e, we need to introduce the notion of the *principal curvatures* κ_1 and κ_2 of a surface. They are the extreme values of normal curvature in a point p of the surface, *i.e.* the curvature of the intersection of a plane containing the normal vector of the surface through p and the surface itself. They measure how the surface bends in different directions at this point. The combination of the two principal curvature gives a useful measure for the curvature of the surface. The *Gaussian curvature g* and the *mean curvature h* are defined as

$$g = \kappa_1\kappa_2, \qquad (4.14)$$

$$h = \frac{\kappa_1 + \kappa_2}{2}. \qquad (4.15)$$

4.2 Minkowski functionals

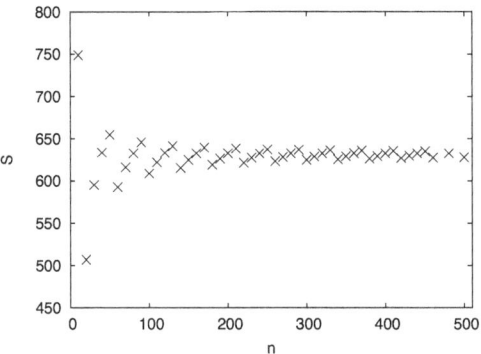

Figure 4.8.: Surface vs. discretization for ideal cylinders.

Now, the integral Gaussian curvature $G(A)$ and the integral mean curvature $H(A)$ are given by respective integration over the surface ∂A:

$$G(A) = \int_{\partial A} \kappa_1 \kappa_2 df, \qquad H(A) = \frac{1}{2} \int_{\partial A} (\kappa_1 + \kappa_2) df. \qquad (4.16)$$

where df is the area element on ∂A.

The mean breadth B is proportional to the integral mean curvature $H(A)$:

$$B(A) = \frac{H(A)}{2\pi}, \qquad (4.17)$$

whereas the Euler characteristic is proportional to the integral Gaussian curvature (Gauss-Bonnet theorem):

$$G(A) = 2\pi \chi_e(\partial A) = 4\pi \chi_e(A). \qquad (4.18)$$

The last part of Equation (4.18) is true for three-dimensional bodies, according to Equation (4.8).

While the Gaussian curvature is a measure for the topology of the surface, the mean curvatures serves more as a quantitative measure. Consider as an example the values for a point on the surface of an ellipsoid: κ_1 and κ_2 both are positive, so the Gaussian curvature as well as the integral Gaussian curvature both will always be positive, which is a characteristic feature of the topology. If we compare this with a point on a locally saddle shaped surface, we obtain a negative Gaussian curvature because κ_1 and κ_2 have opposite signs. For the mean curvature, in contrast, the result depends on the quantitative values of the principal curvatures. Therefore we expect the Euler characteristic to be a much more powerful observable when it comes to the decision which mesophase is present, where we are mainly interested in the topology of the structures.

Usually, the mean curvature is known in literature [57, 58] as an instrument to distinguish between lamellae and gyroid phase on the one hand and cylinders and spheres on the other hand, because the former structures have so-called minimal surfaces: the mean curvature equals zero everywhere on the surface (implying that at *every* point $\kappa_1 = -\kappa_2$, in the case of the lamellae $\kappa_1 = \kappa_2 = 0$), while the surfaces of the latter have a non-zero constant mean curvature. Note, however, that we do not use periodic boundary conditions in order to be fully consistent with our approach of quantifying the visual process of phase determination, where one also just looks at the configuration snapshot. Hence, the value of B resulting from our calculations is not meaningful in the sense of minimal surfaces, since the resulting mean curvature will always be that of the whole surface of the hexahedron that represents the part of the lamella inside of the simulation box, and this surface can never be minimal.

We conclude that among the Minkowski functionals, the only interesting one for our approach to block copolymer mesophase characterization is the Euler characteristic, hence we will focus in the following on the investigation of the fourth Minkowski functional, χ_e.

4.2.3.1. Characterization of block copolymer mesophases using the Euler characteristic

As previously illustrated, the Euler characteristic describes the topology of the structures in the three-dimensional image. χ_e equals the number of regions of connected black voxels plus the number of enclosed white regions (or cavities) minus the number of tunnels of white voxels piercing through the regions of connected black voxels.

Thus, for the topologies appearing in block copolymer mesophases, one should expect the following values of χ_e (always looking at the minority component, *i.e.* setting the B voxels to "black"): due to the fact that we limit the calculation to the configuration inside of the simulation box, and do not use periodic boundary conditions, in the lamellar phase,

4.2 Minkowski functionals

χ_e should equal the number of lamellae in the box, just as in the cylindrical or spherical phase it will amount to the number of cylinders or spheres, respectively. In the perforated lamellar phase however, we expect χ_e to equal the number of minority lamellae minus the total number of holes perforating the lamellae. Thus, χ_e should be negative, if on average there is more than one hole in every perforated layer.

In the same spirit, we expect a negative value for the gyroid phase, because its structure forms a percolating mesh.

Hence, the Euler characteristic of the minority component is able to differentiate between at least the lamellar, cylindrical and spherical phases on the one hand and the perforated lamellar and gyroid phases on the other hand just by the difference in sign. Furthermore, we expect χ_e to have a higher value for the spherical phase compared with the cylindrical phase, and for the latter a higher value than for the lamellar phase, because of the number of respective objects in the simulation box.

CHAPTER 5

Bulk simulations

In this chapter, after an introduction to the simulation method, the results of bulk simulations using the model that has been introduced in chapter 3 will be presented.
In an exemplary manner, the determination of the phase diagram will be described, and other observables that we calculated from the results of the simulations will be discussed. Parts of the work presented in this chapter have already been published in *Soft Matter* [59].

5.1. Simulation method

This section will give an overview of technical details of the simulation method.

5.1.1. The Monte Carlo Method in the canonical ensemble

Monte Carlo (or short MC) methods are a class of algorithms that rely on repeated random sampling to solve mathematical or physical problems, justified by the law of great numbers. They are typically applied in cases where an exact analytical solution is too complex or impossible. There is a large number of Monte Carlo approaches, widely-used in many areas of physics. However, we will focus in the following on the method employed in this work, *i.e.* the *Metropolis algorithm*.

The Metropolis Algorithm

The Metropolis algorithm [60] is a Monte Carlo method that generates Boltzmann-distributed phase space samples. The obtained states form a *Markov chain*, i.e. state $n+1$ depends on state n only.

The Boltzmann distribution is the distribution of the states in the canonical ensemble. It gives the probability p_j to find the system in state j:

$$p_j = \frac{1}{Z} e^{-\beta E_j} \tag{5.1}$$

with the "inverse temperature" $\beta = 1/k_B T$, with k_B being the Boltzmann constant, E_j the energy of state j, and the normalizing constant

$$Z = \sum_{k=0}^{\infty} e^{-\beta E_k}, \tag{5.2}$$

which is also called *partition function* of the canonical ensemble.

Monte Carlo steps consist of trial moves in phase space from state $\mathbf{x}_n$ to state $\mathbf{x}_{n+1}$. Trial moves are generated by picking, at random, one particle out of the box. Then, this particle is attempted to be moved by a vector of random length and random direction.

The energy difference between both points in phase space

$$\Delta E = E(\mathbf{x}_{n+1}) - E(\mathbf{x}_n) \tag{5.3}$$

is calculated and the new configuration is accepted with the probability

$$A_{n \to n+1} = \min\left(1, \exp\left(-\frac{\Delta E}{k_B T}\right)\right). \tag{5.4}$$

This probability is also called *acceptance rate* and is realized by generating a random number in the interval $[0,1]$ and accepting the move if the number is smaller than $A_{n \to n+1}$. This means if $\Delta E \leq 0$, the move is always accepted, and if $\Delta E > 0$, the move is accepted with Boltzmann probability.

Properties of the system that one is interested in are calculated by averaging over samples, or their full distribution is considered.

5.1 Simulation method

Detailed balance

A crucial property of equilibrium is that the probability of the system leaving a given state must be exactly equal to the probability that the system reaches that state leaving any other state. If P_i is the probability of the system being in state i, and $A_{i \to j}$ is the probability of the system going from state i to state j, then this equilibrium condition translates to:

$$\sum_j P_i A_{i \to j} = \sum_j P_j A_{j \to i}. \qquad (5.5)$$

This condition is fulfilled by all means when the much stronger condition of *detailed balance* is satisfied:

$$P_i A_{i \to j} = P_j A_{j \to i}. \qquad (5.6)$$

To produce results without an unknown bias, it is sufficient to take care that this condition is satisfied, meaning that the probability of each move has to be equal to the probability of the reverse move.

It can easily be seen that Equation (5.4) obeys the condition in Equation (5.6).

Periodic boundary conditions

The number of degrees of freedom that can be handled in computer simulations is far away from the thermodynamic limit. To avoid that the system boundaries affect the results too much, the common procedure is to employ periodic boundary conditions that mimic the presence of an infinite bulk surrounding the model system. This is done by reiterating periodically the simulated system by images of itself. A particle that leaves the simulation box on one side reenters the box on the other side.

Although the use of periodic boundary conditions proves to be very efficient to simulate bulk systems, one has to be aware of the fact that such boundary conditions can lead to artifacts that are not present in a truly macroscopic system, *e.g.* long wavelength phenomena may be disrupted, and wrong correlations can be produced.

Saving CPU time

The energy calculation between non-bonded interaction partners is the most time-consuming part of molecular simulations. Even with a pair potential with a finite interaction range as it is the case in our model, one still has to evaluate all pair distances. In a system with N particles, there are $N(N-1)/2$ particle pairs. This implies that the time which is needed for the energy calculation scales as N^2.

To save computing time, we use an algorithm that is known as the *cell-list method* [61]: the simulation box is subdivided into sub-boxes with a size equal or slightly larger than the maximum interaction radius of non-bonded interactions. A particle that is located in a given sub-box can only interact with particles in the same or neighboring sub-boxes. After each Monte Carlo move, the sub-box information has to be updated. The time for the allocation of the particles to the sub-boxes scales with N, and the total number of sub-boxes that has to be considered for the calculation of the interaction is independent of the system size. Hence the use of sub-boxes allows us to reduce the time that is needed for the energy calculation to scale as N.

5.1.2. Simulation details

The soft-tetramer model which we have introduced in chapter 3 has been studied using Monte Carlo simulations in the canonical ensemble, *i.e.* the temperature T, the volume V and the number of chains N_c were held constant.

Two combinations of the parameters V and N that have been studied for a wide range of incompatibility χN and diblock composition f will be discussed here:

- $N_c = 1728$, $V = L_{box}^3 = 1728\sigma^2$, linear box size $L_{box} = 12$, chain density $\rho_c = 1$,
- $N_c = 1728$, $V = 1331\sigma^2$, linear box size $L_{box} = 11$, chain density $\rho_c = 1.3$.

Both systems were simulated using a cubic box with periodic boundary conditions.

For these systems, we can estimate the invariant degree of polymerization (*cf.* chapter 3.3) to

$$N_{\text{inv}} = 216\rho_c^2 = \begin{cases} 216 & \rho_c = 1, \\ 365 & \rho_c = 1.3. \end{cases} \quad (5.7)$$

These values are in the range that is typically reached for molecular simulations. However, they are outranged by the values that are reached in experiments by some orders of magnitude.

The initial configuration is generated by placing the spheres into the box according to an ordered lattice pattern. Starting from this configuration, 1000 Monte Carlo steps are

5.2 Phase diagram

carried out at temperature $T = 1$, which corresponds to the completely symmetric case, $\chi = 0$. The mean squared displacement of the spheres after 1000 Monte Carlo steps is in the order of L_{box}, so we can safely assume that this "annealing" time is sufficient to bring the system into a completely isotropic state.

One Monte Carlo step consists of $N_s = 4N_c$ trial moves of randomly chosen spheres (where N_s is the total number of spheres), *i.e.* on average every sphere is attempted to be moved once in every Monte Carlo step by a vector of random direction and a length randomly distributed between zero and 0.4σ. This maximum step size has been adapted to obtain an average acceptance rate around 50% which is usually considered to be the optimal value in Monte Carlo simulations [62]. The value varies depending on χN.

The number of Monte Carlo steps required to reach equilibrium ranges from $4 \cdot 10^4$ to more than $4 \cdot 10^6$. Typically, another $2 \cdot 10^6$ steps are employed for equilibrium sampling.

In order to get uncorrelated configurations for ensemble averaging, we monitor the system properties with a sampling rate of $(1000 \text{ MC steps})^{-1}$.

We have also tried to reach faster equilibration in the (μ, V, T) ensemble by employing grand-canonical Monte Carlo moves, however this did not prove to be useful. For detailed information refer to Appendix B.

5.2. Phase diagram

5.2.1. Examples of characterization

By means of the method that we have introduced in chapter 4.1, we identify clusters of each component and use specific properties of these clusters like their shape and number to determine the phase in question. Giving a typical example for each identified mesophase, we will illustrate the procedure of characterization in the following.

Lamellar phase

In Figure 5.2, an example of lamellar phase characterization is shown for the parameters $f = 0.66$, $\chi N = 30$, $\rho_c = 1.3$. A corresponding simulation snapshot is shown in Figure 5.1. As we expect for the lamellar phase, we find about the same number of majority (A) and minority (B) clusters, when we look at the cluster number distributions in Figure 5.2(a).

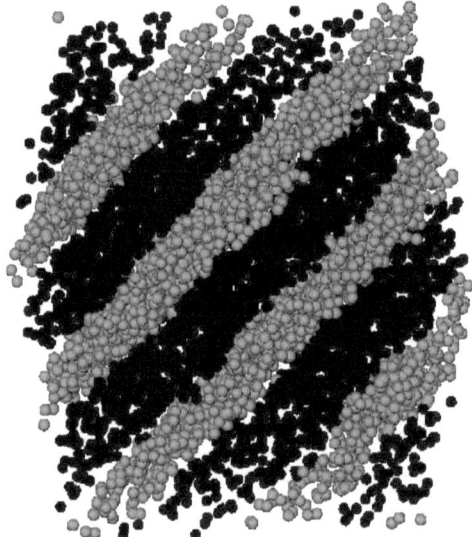

Figure 5.1.: Lamellar (L) phase of the soft-tetramer model observed at $f = 0.66$ and $\chi N = 30$ ($L = 11$, $\rho_c = 1$). In all representations of simulation snapshots, the spheres are depicted with reduced diameter for better visibility.

5.2 Phase diagram

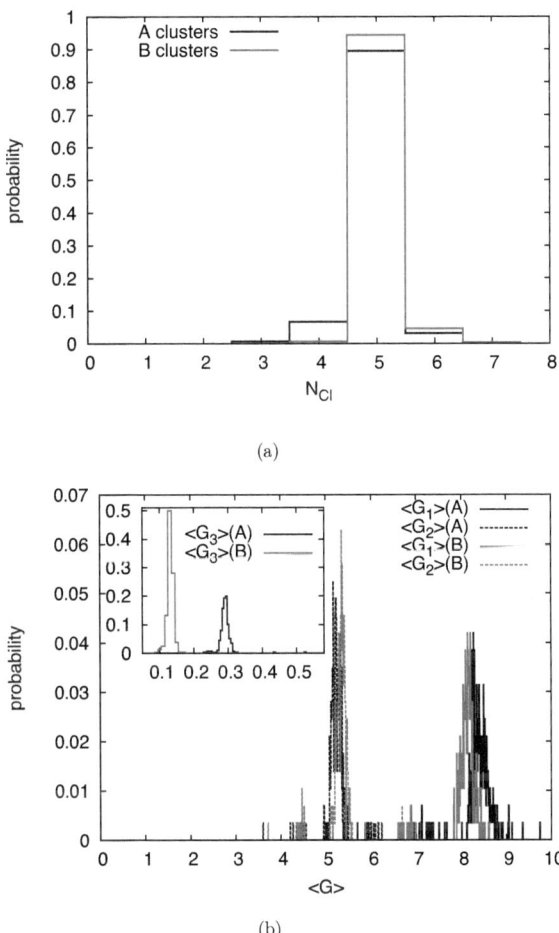

Figure 5.2.: Cluster number distributions for majority (A) and minority (B) phases for a lamellar mesophase for $\rho_c = 1.3$, $f = 0.66$ and $\chi N = 30$ (a), and distributions for the three mean eigenvalues of the gyration tensor of the clusters in (b). For better visibility, the smallest eigenvalues $\langle G_3 \rangle$ are shown in an inset with a different scale.

Each component builds about five clusters. Looking at the snapshot in Figure 5.1, we can confirm these findings simply "by the eye".

The gyration tensor eigenvalues G_i ($i \in \{1,2,3\}$) with $G_1 > G_2 > G_3$ are calculated for each identified cluster and then averaged over all clusters of the same component in every configuration sample. The distributions of the averaged values $\langle G_i \rangle$ are shown in Figure 5.2(b).

The first and second eigenvalues, respectively, amount each for both components approximately to the same values (*i.e.* $\langle G_{1,A} \rangle \approx \langle G_{1,B} \rangle$ and $\langle G_{2,A} \rangle \approx \langle G_{2,B} \rangle$), as it is expected for the lamellar phase. This is due to the fact that in contrast to the third eigenvalue, the first and second ones do not depend on the composition f but on the box size and the angle that the lamellae draw with the box sides, and these conditions are the same for both components.

The eigenvalues of the gyration tensor relate to the dimensions of the cluster as shown in Equation (4.3). In the given example, the values of the first and second eigenvalues indicate that the lamellae lie tilted in the box: If the lamellae lay parallel to the box sides, the two long side lengths of each lamellar cluster should be equal to the linear box size L_{box}, so that all clusters in one configuration should have $G_1 = G_2 = L_{box}^2/12 \approx 10$ (*cf.* Equation (4.3)), and thus also $\langle G_1 \rangle = \langle G_2 \rangle = 10$. However, if the lamellae lie diagonally in the box, the layers distant from the center of the box have significantly smaller side lengths and contribute less to the average value, so that $G_1, G_2 < 10$ which is the case in our example. Furthermore, the fact that the second eigenvalue is considerably lower than the first one indicates that the lamellae are not exactly in the diagonal of the box, but are tilted by a different angle in two directions, so that one of the long sides of the cuboid-shaped cluster is longer than the other one.

The distributions of the smallest eigenvalues of both components (shown in the inset of Figure 5.2(b)) show very narrow peaks. They correspond to the thickness of the lamellae. Because of $f > 0.5$, the $\langle G_{3,A} \rangle$ of the majority component has a higher value (about $0.29\sigma^2$, corresponding to a lamella thickness of $d_A \approx 1.9\sigma$) than that of the minority component (about $0.14\sigma^2$, corresponding to a lamella thickness of $d_B \approx 1.3\sigma$).

If we compare the lamella spacing calculated in this way with the snapshot in Figure 5.1, we find that it is still smaller than what is visible in the snapshot. The reason for this discrepancy is that our method only takes into account the centers of the spheres. This effect is expected to be more important for smaller lengths. For lengths in the order of the linear box size, we do not see this effect, as we will see for the eigenvalue distributions

5.2 Phase diagram

of the majority clusters in the following sections describing the perforated lamellae phase and the cylinder phase.

To get a better picture of the true size of the lamellar clusters, we apply the cluster algorithm on configurations that we mapped to $50 \times 50 \times 50$ cubic cells with the method that has been described in chapter 4.1.1. We discuss the procedure and the results in chapter 5.3.

However, since we do not use the absolute thickness of the lamellar layers for the identification of the phases, we can leave this issue unregarded for the moment and carry on with the center-of-sphere based cluster analysis.

Perforated lamellae phase

As an example for the perforated lamellae (PL) mesophase, Figure 5.3(a) shows a snapshot at parameters $\rho_c = 1.3$, $f = 0.72$ and $\chi N = 21$. A single perforated lamella is shown in Figure 5.3(b).

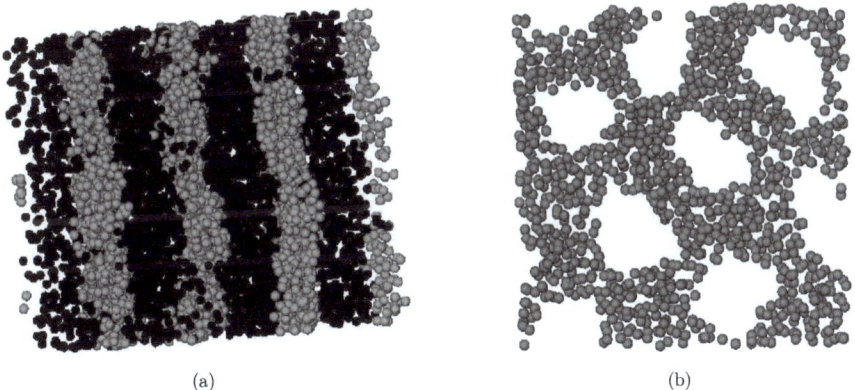

(a) (b)

Figure 5.3.: Snapshot of the perforated lamella (PL) phase at $\rho_c = 1.3$, $f = 0.72$ and $\chi N = 21$ (a) and section showing just one perforated minority lamella (b).

The distribution of cluster numbers in Figure 5.4(a) shows that we typically find only

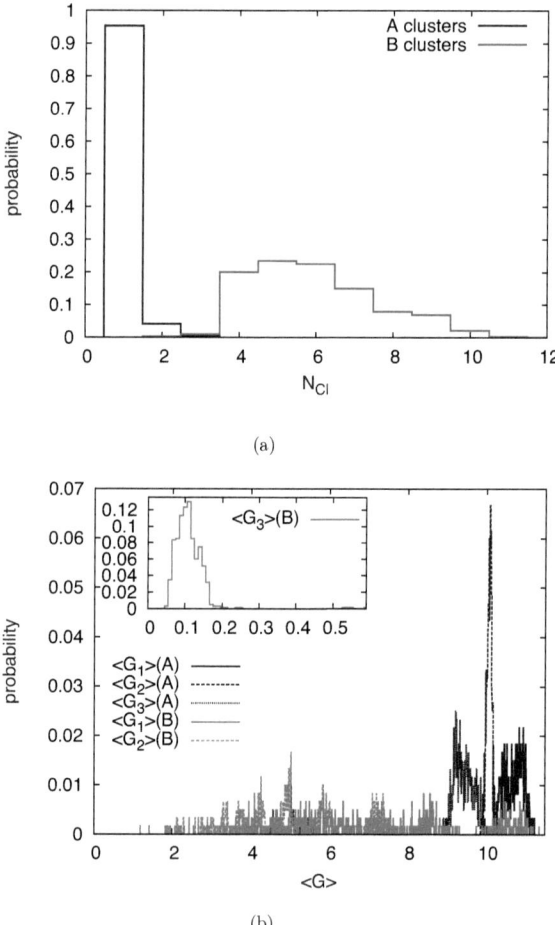

Figure 5.4.: Cluster number distributions for majority (A) and minority (B) components for the perforated lamella mesophase at $\rho_c = 1.3$, $f = 0.72$ and $\chi N = 21$ (a) and distributions for the three eigenvalues of the gyration tensor of the clusters (b).

5.2 Phase diagram

one majority (A) cluster in the PL phase which extends over all of the simulation box, as can be seen in Figure 5.4(b). All three eigenvalues of the majority cluster gyration tensor approximately amount to 10, corresponding to a cluster side length of $L_{cl} = 11$ according to Equation (4.3), which corresponds to the side length L_{box} of the cubic simulation box. The number distribution of minority clusters peaks at the same value as for the lamellar phase, but has broadened due to fluctuation effects which can break down one perforated layer into more than one cluster. These fluctuation effects are also visible in the distribution of $\langle G_{1,B} \rangle$ and $\langle G_{2,B} \rangle$ for the minority clusters. They now extend over a broad range reaching down to about $2\sigma^2$, which corresponds to a decayed cluster length of less than 5σ, i.e. less than half of the linear simulation box size.

Only the smallest minority eigenvalue (shown in the inset of Figure 5.4(b)) has a very narrow distribution that peaks at $\langle G_{3,B} \rangle \approx 0.12$, corresponding to a lamella thickness of $d_B = 1.2\sigma$.

Cylinder phase

As a final example we now turn to the cylinder (C) phase, since we did not find a gyroid nor a sphere phase. As an example we show the structure that forms at parameters $\rho_c = 1.3$, $f = 0.81$, $\chi N = 48$. The simulation snapshot in Figure 5.5(a) shows nicely the hexagonal arrangement of the cylinders formed by the minority component. Figure 5.5(b) shows a side view of the minority component only, with the majority segments removed from the image, in order to show the cylindrical character of the structure.

Similar to the perforated lamellar phase, the majority component consists of just one cluster extending over the whole simulation box, as the distributions of cluster numbers and eigenvalues for the majority component in figures 5.6(a) and 5.6(b) show: Again, $\langle G_{i,A} \rangle \approx 10$ for $i \in \{1, 2, 3\}$. Here, the distributions of $\langle G_{i,A} \rangle$ are even narrower than for the majority cluster in the perforated lamellar phase. We can understand this if we look at the cluster number distributions: For the perforated lamellae phase, the majority cluster at times decays into two or even three clusters, which leads to a broadened distribution for the averaged eigenvalues. For the cylinder phase in contrast, this basically never happens. Turning to the minority component, we find about 18 clusters with two small and one larger eigenvalue, indicating a cylindrical shape. The cylinders lie tilted in the cubic simulation box as can be seen in Figure 5.5(b), hence, the mean largest eigenvalue $\langle G_{1,B} \rangle$ is

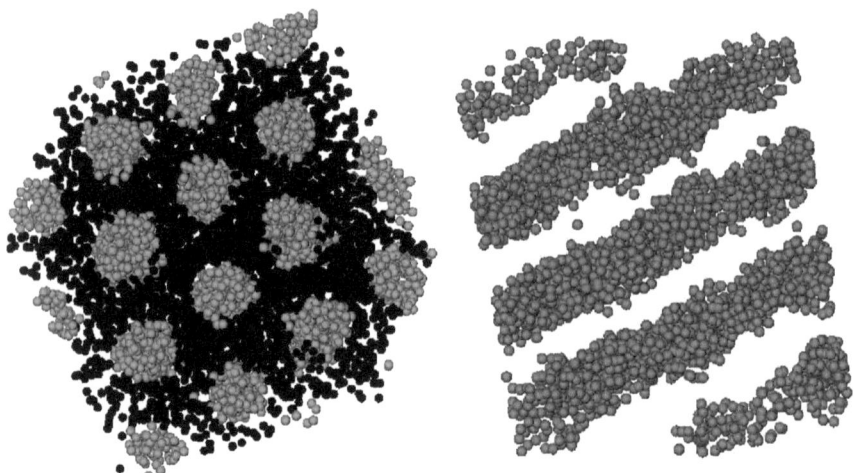

(a) Head-on view onto the plane perpendicular to the cylinders.

(b) Side view, only B blocks are shown.

Figure 5.5.: Simulation snapshot of the cylinder phase at $\rho_c = 1.3$, $f = 0.81$ and $\chi N = 48$.

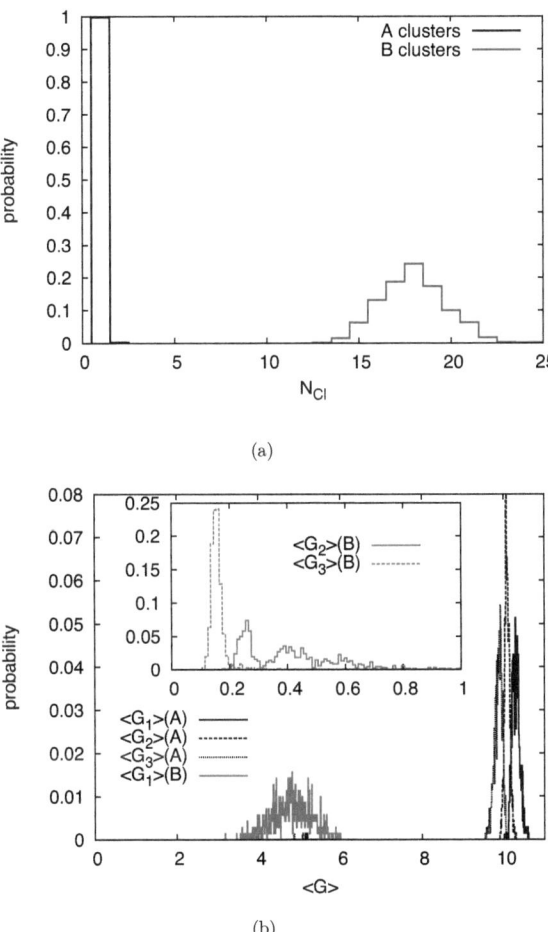

Figure 5.6.: Cluster number distributions for majority (A) and minority (B) phases for the cylinder mesophase at $\rho_c = 1.3$ (a) and distributions for the three eigenvalues of the gyration tensor of the clusters (b).

averaged over cylinders with different lengths and thus smaller than 10, as Figure 5.6(b) shows. The inset of this figure displays the two smaller mean eigenvalues of the minority component. They are comparable in size, but $\langle G_{2,B} \rangle$ clearly has a little higher value than $\langle G_{3,B} \rangle$. This can be caused by a slightly elliptical instantaneous shape of the cylinders' cross section as well as by cylinders being cut longitudinally by the box boundaries.

5.2.2. Phase diagram: results and discussion

Phase diagrams have been obtained in the way that we have exemplified in the last section for some of the phase points. They are shown for both densities separately in Figures 5.2.2 and 5.2.2.

The order-disorder transition for symmetric composition occurs at $(\chi N)_c \approx 15$ for the $\rho_c = 1$ system and at $(\chi N)_c \approx 8$ for the $\rho_c = 1.3$ system. These critical values do not scale linearly with the density, moreover, there seems to be no trivial dependence of the phase diagrams on density at all. Also qualitatively, the two systems differ in that we did not find a cylindrical phase in the low density system for incompatibilities up to $\chi N = 60$.

For the smaller density, fluctuation effects play a larger role, because the invariant degree of polymerization N_{inv} depends on the density (see section 3.3). The lower N_{inv}, the more the system is affected by fluctuations, so a possible explanation could be that the cylinder phase is disrupted by fluctuations in the low density system and transformed into a part of the disordered region.

At $\rho_c = 1.3$, we find transitions from the lamellar phase via a perforated lamellar phase into the cylindrical phase at low incompatibility, whereas at high incompatibilities, there is a direct transition from lamellar to cylindrical phase.

Comparison with phase diagrams from theory, experiment and molecular simulation

We observe direct transitions from the disordered phase into the perforated lamellar and cylindrical phase for $f > 0.5$. This is reported from experiments and molecular simulations, but has not been seen in self-consistent field theory due to the neglected fluctuations, as has been described in chapter 2.2.3.

5.2 Phase diagram

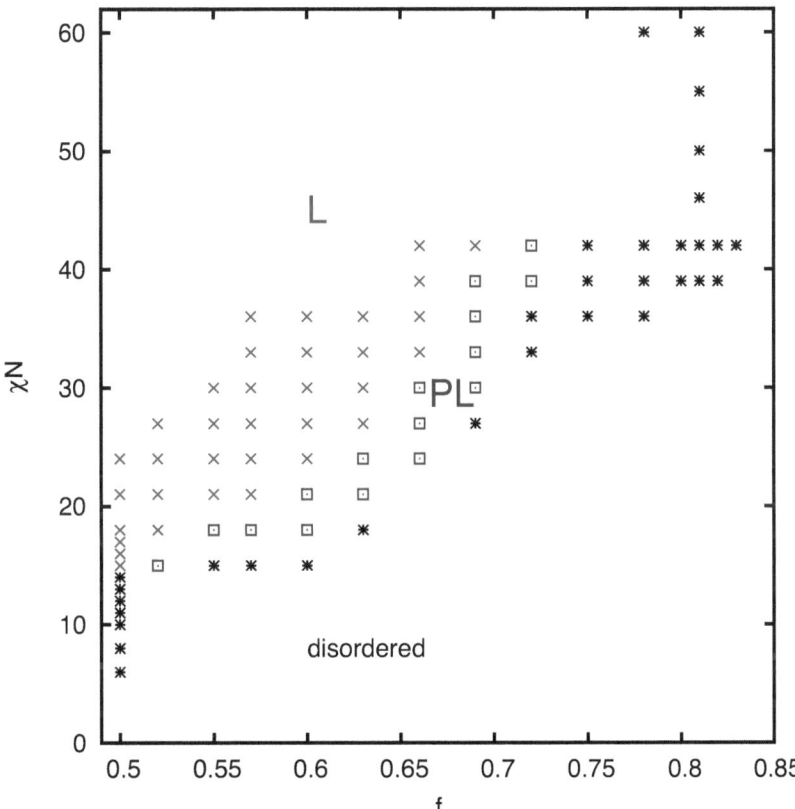

Figure 5.7.: Phase diagram of the soft-tetramer model for box size $L = 12$, chain number density $\rho_c = 1$.

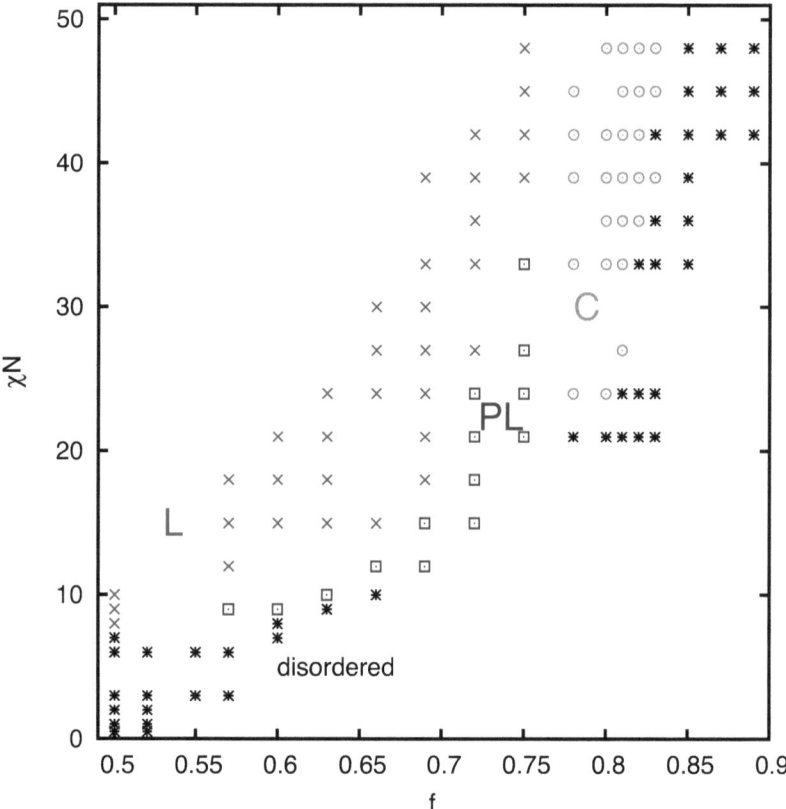

Figure 5.8.: Phase diagram of the soft-tetramer model for box size $L = 11$, chain number density $\rho_c = 1.3$.

5.2 Phase diagram

In neither of the two systems we found a spherical or a gyroid phase. Again, a possible reason could be for both "missing" phases that they are destroyed by fluctuations.
However there is another lucid explanation: Structures with periodicity in one or two directions (e.g. the lamellar and cylindrical phase, respectively) can accommodate themselves corresponding to the periodic boundary conditions by rotating within the simulation box. The gyroid phase, by contrast, is periodic in three dimensions. Its formation is hence especially sensitive to the selection of simulation box size, meaning that it can only be found if the simulation box size is exactly commensurate with the gyroid lattice spacing. For inappropriate sizes of the simulation box, the gyroid phase becomes severely frustrated. When this frustration is present, other phases, in particular the perforated lamellae phase, may be artificially stabilized in lieu of the gyroid phase [63].

Coarse-grained molecular simulations studies previously have also observed only the lamellae, cylinder and perforated lamellae phases, but they have been limited so far to the determination of the order-disorder transition [40] (cf. Figure 2.4). Also on a similar scale as our model, investigations have not reached beyond the determination of the order-disorder transition at symmetric composition [21].
With our simple model, in contrast, we are able to go further and thoroughly determine the equilibrium morphology at a broad range of points in the control parameter space. Moreover, we can investigate phase transitions between ordered phases. This will be discussed in more detail in following sections. Hence, the knowledge that we are able to obtain about the phase behavior is much more complete than in previous molecular simulations.

Simulation box size effects on the phase transition

To verify that the differences between both investigated systems in the position of the order-disorder transition really depend on the density rather than on the size of the simulation box, we also investigate a system at density $\rho_c = 1$ and simulation box side length $L = 11$ for an example value of diblock composition, $f = 0.57$.
Figure 5.9 shows the values of the shape parameters K_1 and K_2 versus incompatibility χN. Both parameters have high values in the disordered state[1] and decrease at the disorder-order transition to their typical values in the lamellar state (cf. section 5.4). The phase transition in the $\rho_c = 1$ system is shifted slightly at best to lower values due to the decreased box size, however this effect is clearly not as important as the density effect as can be seen

---

[1] For the properties of cluster shapes in the disordered phase cf. Appendix A.

if one compares to the $\rho_c = 1.3$ system.

A systematic investigation of finite size effects in Monte Carlo simulations of microphase separation in block copolymer melts is a subtle issue [64].

5.3. Cluster algorithm on discretized configurations

The cluster analysis algorithm has also been applied to images mapped to $50 \times 50 \times 50$ cubic A and B voxels by the same discretization method that has been applied for the calculation of Minkowski functionals and has been described in chapter 4.1.1. The objective was to analyze clusters with this higher order discretization to get a more realistic picture of the cluster dimensions than by applying the original method that only uses the comparatively low resolution on the center-of-sphere level.

Here, the condition which is used in the original algorithm (*cf.* chapter 4.1.1) that the cluster has to consist of more than three spheres to be taken into account is translated into the condition that a cluster has to consist of more than 54 voxels to be taken into account, because in the simulation box we have 6912 spheres that are mapped onto 125000 voxels, so on average three spheres correspond to approximately 54 voxels.

The averaged eigenvalues of the clusters resulting from the application of the cluster algorithm to the discretized configurations are compared in Figure 5.10 with those obtained with the "direct method" as described in the previous sections.

Whereas $\langle G_1 \rangle$ and $\langle G_2 \rangle$ remain more or less the same compared with our usual soft sphere center method as Figure 5.10(a) shows, $\langle G_{3,B} \rangle$ is shifted a tiny bit and $\langle G_{3,A} \rangle$ considerably to higher values (cf. Figure 5.10(b)).

$$\langle G_{3,A} \rangle_{mapped} \approx 0.43 \longrightarrow L_{3,A} = \sqrt{12 \cdot \langle G_{3,A} \rangle_{mapped}} \approx 2.27$$

$$\langle G_{3,B} \rangle_{mapped} \approx 0.15 \longrightarrow L_{3,B} = \sqrt{12 \cdot \langle G_{3,B} \rangle_{mapped}} \approx 1.34$$

A case where this method can be employed beneficially to determine the bulk lamellar period will be discussed in chapter 6.3.

5.3 Cluster algorithm on discretized configurations

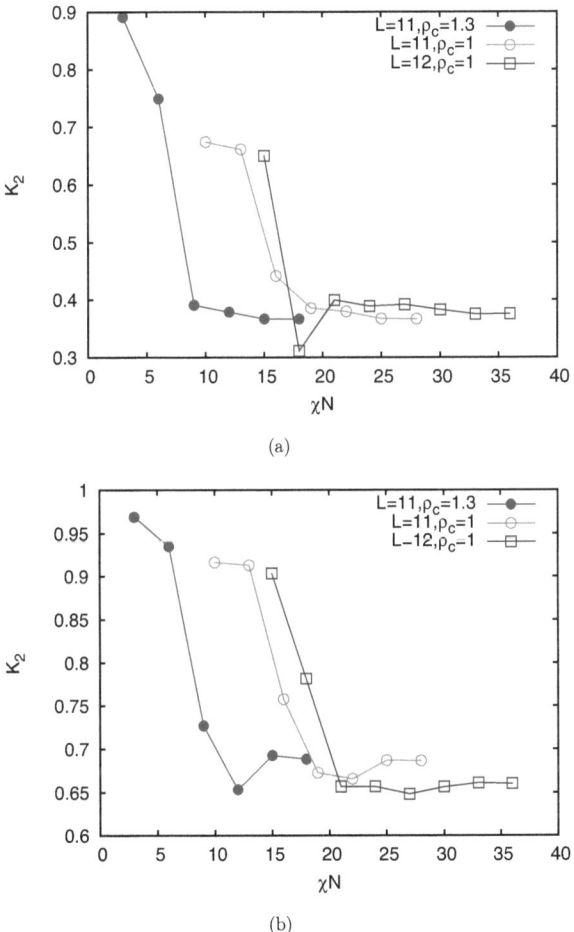

Figure 5.9.: Dependence of the location of the disorder-order transition on box length and density, shown examplarily for the shape parameters K_1 (a) and K_2 (b): The phase transition is shifted slightly to lower values of χN when the simulation box size is decreased, however this is effect is not as important as the shift due to the higher density.

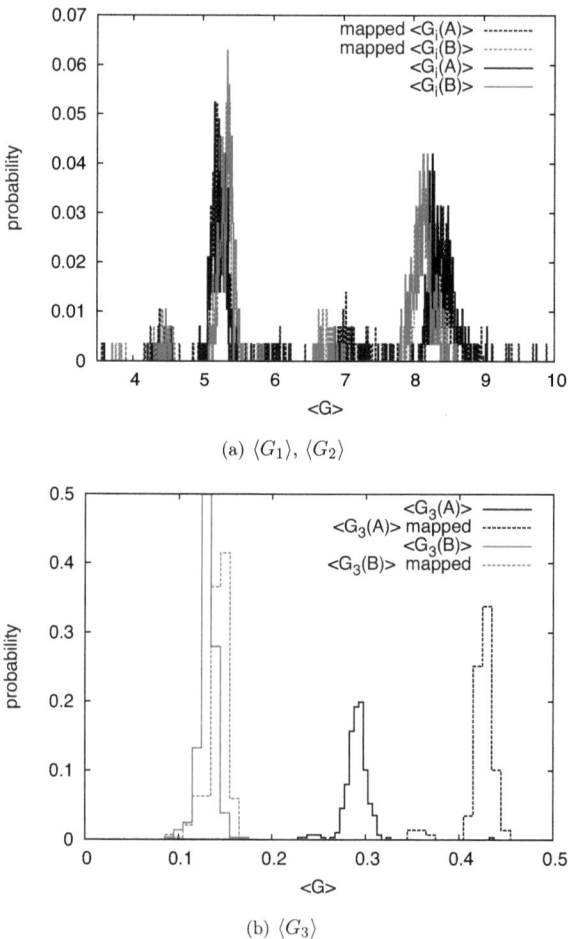

Figure 5.10.: Results of the cluster algorithm applied on discretized configurations (dotted lines) compared with cluster algorithm applied directly to configuration (solid lines) for L=11, f=0.66, $\chi N = 30$, $\rho = 1.3$.

5.4 Shape parameters

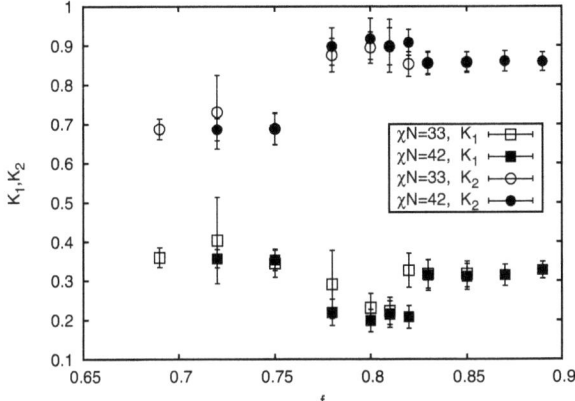

Figure 5.11.: Composition dependence of the shape parameters K_1 (squares) and K_2 (circles) for $\chi N = 33$ (open symbols) and $\chi N = 42$ (closed symbols) in the regime of the ordered mesophases, $\rho_c = 1.3$.

5.4. Shape parameters

Two additional observables that can directly be calculated from the results of the cluster algorithm are the shape parameters K_1 and K_2 that were defined in section 5.4. They can be used as supplementary criteria for the determination of the phase transitions between the ordered phases. We calculated K_1 and K_2 based on the gyration tensor eigenvalues of the minority component clusters that have been determined as described in section 4.1.1. In the lamellar phase, K_1 and K_2 should be both approximately equal to 1/2, because point $(K_1 = 1/2, K_2 = 1/2)$ in the (K_1, K_2) plane is attributed to the ideal disk. In the cylinder phase, K_1 should be close to zero and K_2 close to one, since point $(K_1 = 0, K_2 = 1)$ represents the ideal rod. For the disordered phase, the values depend on the cluster size and shape fluctuations occurring within this phase. Hence we are not able to make any predictions about the shape parameters in the disordered state.

In Figure 5.11, we show the values for K_1 and K_2 at density $\rho_c = 1.3$ for two different values of χN above the order-disorder transition as a function of the composition f. The order-order transition from the lamellar to the cylindrical phase at about $f = 0.77$ is clearly visible, just as well as the transition from the cylindrical to the disordered phase around $f = 0.82$. Compare also to Figure 5.2.2.

Figure 5.12 compares the values of ideal shapes with the values we found in our simulations.

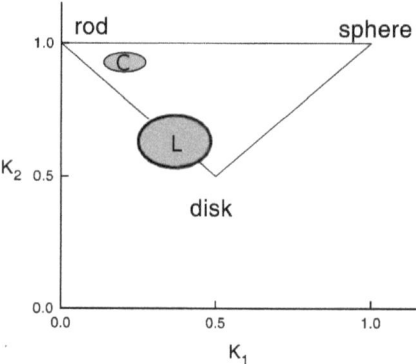

Figure 5.12.: The gray shaded areas indicate the regions in the (K_1, K_2) plane where the simulated lamellar (L) and cylindrical (C) structures are located.

5.5. Euler characteristic

We calculated the Euler characteristic of the minority component mapped to $50 \times 50 \times 50$ cubic B voxels. See section 4.2 for a detailed description of the procedure.

The Euler characteristic of the minority component as a function of chain composition is shown in Figure 5.13(a) for the system with lower density $\rho_c = 1$ for several values of incompatibility χN. The phase transition from lamellar to perforated lamellar phase is visible as the Euler characteristics goes from positive to negative values. This phase transition occurs later in f for higher incompatibility χN: For $\chi N = 21$, χ_e becomes negative at $f = 0.6$, for $\chi N = 24$ at $f = 0.63$, for $\chi N = 27$ and $\chi N = 30$ at $f = 0.66$ and so on. This is consistent with the phase diagram in Figure 5.2.2.

For better visibility, this picture shows only the ordered regime; for the behavior of the Euler characteristic in the disordered phase, the highest three values of χN are shown again in Figure 5.13(b) for the whole range of composition f. Since the number of unconnected minority structures increases in the disordered phase with increasing f, χ_e also increases. For $f = 0.75$ (and also $f = 0.72$ at $\chi N = 36$), the Euler characteristic is still negative although these points have been previously characterized as disordered, using the cluster identification method. The topology of the structures in the disordered phase is not known a priori, so it could be possible that the structures are shaped in a way that they result in negative Euler characteristics. Possibly, however, this could also be indicating that there might still be perforated lamellae states intermixed.

In Figure 5.13(c), the Euler characteristic is plotted against composition for the higher density system for two values of χN. At density $\rho_c = 1.3$, we have already discussed that there is a direct transition from lamellar to cylindrical phase for high values of incompatibility, whereas at smaller χN the perforated lamellar phase appears in between. As in the case described before, the Euler characteristic turns negative for the perforated lamella phase, for $\chi N = 24$ this happens at $f = 0.72$ and $f = 0.75$. Then there is a plateau regime between $f = 0.78$ and $f = 0.83$ indicating the cylinder phase, before χ_e increases upon entering the disordered phase. At $\chi N = 42$, the direct transition from the lamellar phase to the cylinder phase is visible. However, for $f = 0.75$, the Euler characteristic goes close to zero, indicating that there might still be perforated lamella states intermixed with the lamellar states at this point.

5.6. Structural details

In the last part of this chapter, we will discuss several observables that can be calculated from the simulation results and that provide a more detailed insight into the structural behavior of the soft-tetramer model.

5.6.1. Chain stretching

The random phase approximation (RPA) for copolymer melts formulated by Leibler in 1980 [35] assumes that one can treat the copolymer chains in the disordered phase as "unperturbed" by interchain interactions, implying simple Gaussian behavior for the conformation of the copolymer chain for temperatures above the order-disorder transition. However, Binder and Fried [65] showed that there exist significant deviations from Gaussian

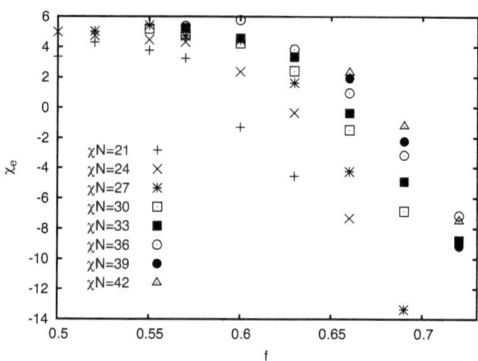

(a) $\rho_c = 1$, ordered regime is shown only.

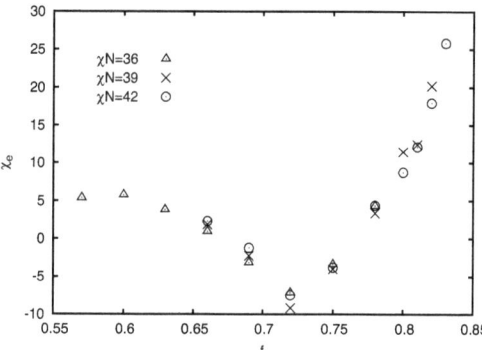

(b) $\rho_c = 1$, all of the investigated composition range is shown.

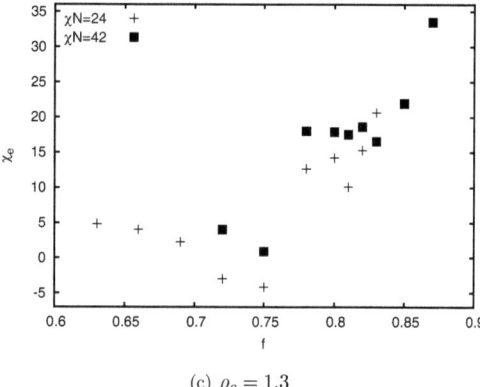

(c) $\rho_c = 1.3$

Figure 5.13.: Composition dependence of the Euler characteristic χ_e of the minority component

5.6 Structural details

behavior, namely a gradual stretching of the chains with increasing interaction strength. They found for a three-dimensional lattice model for a symmetric diblock copolymer melt that the average separation between the centers of mass of the two blocks, R_{AB}, increased strongly with decreasing temperature. This effect has also been shown in experiments [66]. To check whether our model can reproduce this effect, we monitor the average separation between the centers of mass of the A block and the B block as a function of incompatibility χN for symmetric composition. This is the absolute value of the vector shown in Figure 5.15(b). It is a well-defined observable in our model and its behavior as a function of χN is shown for both investigated densities in Figure 5.14.

We observe a clear stretching effect of the diblock chains approaching the order-disorder transition (located at $(\chi N)_c \approx 15$ at $\rho_c = 1$, and $(\chi N)_c \approx 8$ at $\rho_c = 1.3$) and also continuing into the ordered phase.

Presumably, our simple model will not correctly reproduce the chain stretching effect in the strong segregation limit, where the chains are stretched so strongly that the behavior of the sub-chains can not be regarded as Gaussian any more. However, in the weak segregation regime for which our model has been constructed, the effect can not only be observed, it also has the same order of magnitude (about 15%) as the effect observed in molecular simulations [65], indicating that the stretching does happen between Gaussian subunits of blocks rather than on a local monomer scale.

5.6.2. Saupe order tensor

In this section, we want to address the question whether we can find an orientational order of the tetramers in the ordered regime. To this end, we use the Saupe order tensor, traditionally used for determining the nematic order of liquid crystals. Liquid crystals consist of aspherical molecules to which an axis of rotational symmetry can be assigned. If those molecules are not aligned in every direction with equal probability, the system is said to have nematic order, and one can use the axis of rotation, also called the *molecule director*, to determine a preferred direction for the system, or the *system director*.

The Saupe order tensor [67] is defined by:

$$Q = \frac{1}{2}\left(\frac{3}{N}\left(\sum_{i=1}^{N}\hat{e}^{(i)} \otimes \hat{e}^{(i)}\right) - \mathbb{1}\right) = \frac{3\left\langle\hat{e}^{(i)} \otimes \hat{e}^{(i)}\right\rangle - \mathbb{1}}{2} \qquad (5.8)$$

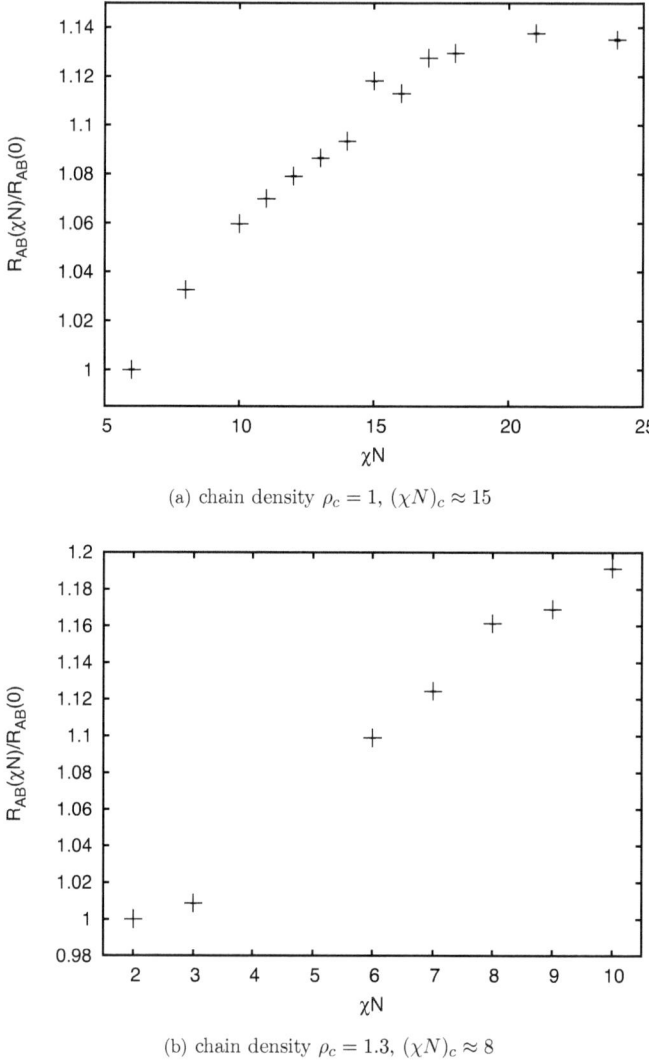

(a) chain density $\rho_c = 1$, $(\chi N)_c \approx 15$

(b) chain density $\rho_c = 1.3$, $(\chi N)_c \approx 8$

Figure 5.14.: Distance between the centers of mass of the A and B blocks as a function of χN at $f = 0.5$, normalized to the limit in the disordered state.

5.6 Structural details

where $\otimes$ signifies the dyadic product and $\hat{e}^{(i)}$ is a unit vector in the direction of the respective vector attached to molecule i. The angle brackets here indicate averaging over the system.

The largest eigenvalue of Q serves as an order parameter, which we call S. When the vectors attached to all molecules point in the same direction, $S = 1$; when the orientation is completely random, $S = 0$.

This definition of the order parameter S is equivalent to the original [68] and more commonly used definition using the second Legendre polynomial:

$$S = \langle P_2(\cos\theta) \rangle = \left\langle \frac{3\cos^2\theta - 1}{2} \right\rangle \qquad (5.9)$$

where θ is the angle between system director and molecule director. However, the former definition has the advantage compared to the latter one that one does not have to know the system director before starting with the calculation of the order parameter [67].

We calculated the orientational order of four different axes associated with the tetramer: $\vec{r}_{AA}$ and $\vec{r}_{BB}$, the vectors connecting the centers of both spheres of an A or B block, respectively (illustrated in Figure 5.15(a)), $\vec{r}_{AB}$, the vector connecting the center of mass of the A bock with the center of mass of the B block (Figure 5.15(b)), and $\vec{r}_{A_2B_1}$, the vector connecting the center of the second A sphere with the center of the first B sphere (Figure 5.15(c)). All vectors are normalized to unit vectors $\hat{e}_{AA}$, $\hat{e}_{BB}$, $\hat{e}_{AB}$ and $\hat{e}_{A_2B_1}$, respectively, in order to calculate the Saupe order tensor as given in Equation (5.8).

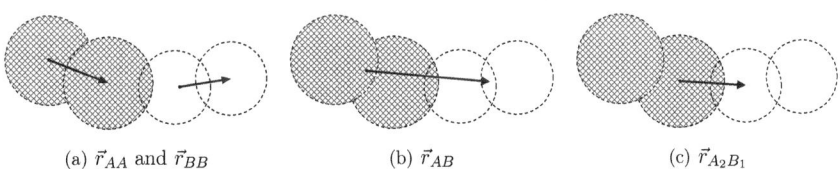

(a) $\vec{r}_{AA}$ and $\vec{r}_{BB}$ (b) $\vec{r}_{AB}$ (c) $\vec{r}_{A_2B_1}$

Figure 5.15.: Molecule directors attached to the tetramer. See text for definitions.

The results of the calculations for symmetric composition are shown in Figure 5.16. The values of S_{AB} and $S_{A_2B_1}$ show approximately the same behavior. Fluctuating around zero in the disordered phase, they rise discontinuously at the phase transition to about $S = 0.3$ and increase further to about $S = 0.5$ as we go deeper into the ordered regime. This also

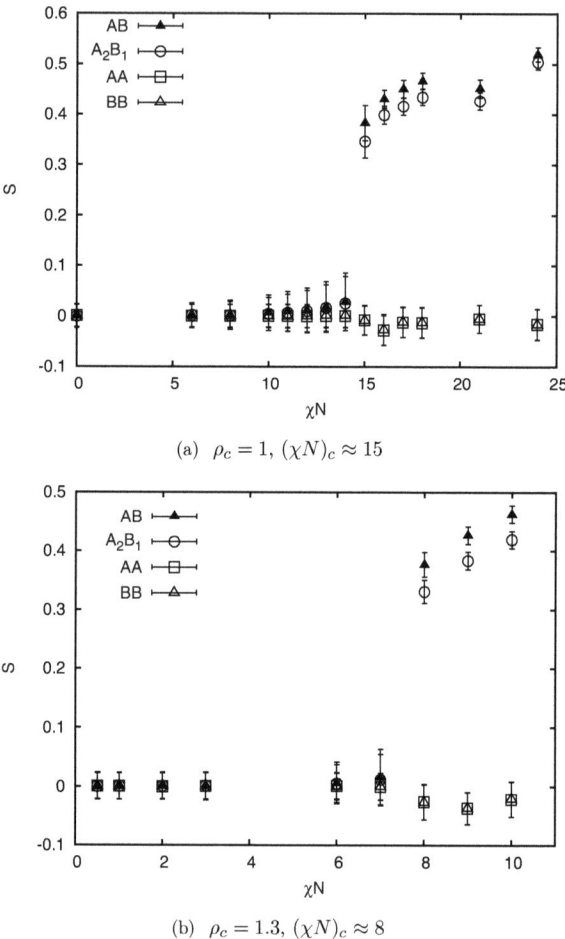

Figure 5.16.: χN dependence of the largest eigenvalue of the Saupe order tensor for all investigated molecule directors, for both systems at symmetric composition $f = 0.5$.

5.6 Structural details

confirms what we found out about the distance R_{AB} in section 5.6.1.
The value for S_{AA} and S_{BB}, however, fluctuate around zero, indicating that there is no ordering within the constituent blocks.

For a nematic liquid crystal domain, this order parameter takes values between 0.3 and 0.8 [68] which is in the same regime as the values we find for S_{AB} and $S_{A_2B_1}$.
There is a clear first-order-like isotropic-nematic transition visible for the vectors $\vec{r}_{AB}$ and $\vec{r}_{A_2B_1}$ at the same values of incompatibility we have already determined for the disordered-ordered transition using the previously described methods.

5.6.3. Lattice site correlation order parameter ψ

Matsen et al. [40] suggested an order parameter for diblock copolymer melt lattice models based on the correlation between lattice sites. We adapted this parameter for our purposes, and as the calculation is only possible on a lattice, we map our configurations first to a $30 \times 30 \times 30$ lattice using the discretization method described in chapter 4.2.2. We evaluate the correlation function

$$G_{ij} = \langle s_i s_j \rangle - \langle s_i \rangle^2, \qquad (5.10)$$

where $s_i = 1$ or -1 if the ith lattice site is an A voxel or a B voxel, respectively, and the angle brackets $\langle \rangle$ here signify thermal averaging (averaging over N_{conf} configuration samples). For each combination of control parameters, the system has been averaged over the same number of configuration samples.
Reference [40] suggests that

$$\psi = \frac{1}{V^2} \sum_{i \neq j} G_{ij}^2 \qquad (5.11)$$

serves the requirements for an order parameter meaning that it should be zero, or at least very small in the disordered phase and large in the ordered phase.
The results for different system densities and incompatibilities are shown in Figure 5.17. In each case, the critical value for $(\chi N)_c$ that had been determined previously could be confirmed by ψ increasing significantly at $(\chi N)_c$.
The phase diagram in Figure 2.4 was established by means of this order parameter [40], but whereas Matsen et al. in their study used heating and cooling runs that produced a hysteresis loop with different values for the phase transition, we are investigating equilibrium states.

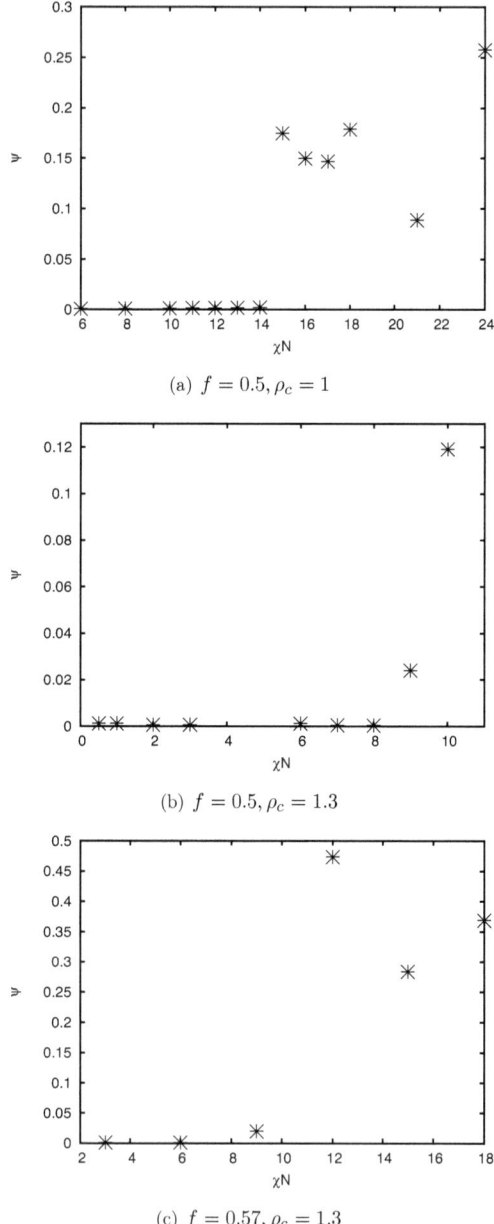

(a) $f = 0.5, \rho_c = 1$

(b) $f = 0.5, \rho_c = 1.3$

(c) $f = 0.57, \rho_c = 1.3$

Figure 5.17.: Correlation order parameter ψ.

CHAPTER 6

Confined polymer melts

In this chapter, we describe the derivation of an effective wall potential for the soft-tetramer model. We carry out molecular dynamics simulations of n-alkane (polyethylene) melts confined between graphite walls to obtain distribution functions from simulations at a more detailed scale than the soft-tetramer model. These simulations will be described in section 6.1. How we use the results of these finer-scale simulations as an input for systematic coarse-graining of the surface interaction will be described in section 6.2. In the last part of this chapter, we will show some first simulation studies of the confined soft-tetramer model.

The work described in this chapter has mostly been done in collaboration with Cheol Jeong and Do Y. Yoon at Seoul National University.

6.1. MD Simulations of confined polyethylene melts

We carry out molecular dynamics simulations[1] of $CH_3(CH_2)_n CH_3$ chains confined between two parallel graphite walls. The aim is to use the resulting center of mass distribution

---

[1] A thorough introduction to the molecular dynamics method can be found in various textbooks [62, 69]. We also give a short outline in Appendix C.

to derive a wall potential for the soft-tetramer model. More exactly, because the effective wall potential has to be valid for the spheres of our tetramer model in which each sphere corresponds to one half of one of the two polymer chains constituting the diblock molecule, we are interested in the distribution of the centers of mass of the half chains, $i.e.$ we look separately for the center of mass of the "subchain" constituted by monomers 1 to $N/2$ and the subchain containing monomers $(N/2+1)$ to N.

We study two systems: a melt of 40 chains with 100 backbone carbon atoms ($n = 98$, "C100"), and a melt of 64 chains with 300 backbone carbon atoms ($n = 298$, "C300"). All molecular dynamics simulations were carried out using GROMACS [70].

6.1.1. Simulation details – polyethylene

We use an "united atom" approach for the n-alkane chains, meaning that every methyl (CH_3) and methylene (CH_2) group is combined into one single effective interaction site with one center of force (see Figure 6.1 for a schematic representation of the united atom approach).

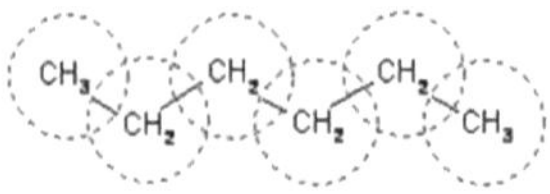

Figure 6.1.: Schematics of the united atom model for alkanes: Each methyl- and methylene group is combined into one single effective particle.

We use an united atom force field for melts of n-alkane chains that has been developed by Wolfgang Paul, Grant Smith and Do Y. Yoon in 1995 [71], and has been widely used and tested since.
The force field consists of two parts: The equations describing the interaction potentials, and the parameters used in these equations. We will specify and explain the equations in the following, whereas for the parameter values we refer the reader to reference [71].
The interaction potentials can be divided in non-bonded interactions, including intermolecular and non-bonded intramolecular interactions, and bonded interactions, comprising

6.1 MD Simulations of confined polyethylene melts

bond-stretching (2-body), angle-bending (3-body) and torsional (4-body) interactions. For atoms that are more than four sites apart along the chain, non-bonded interactions are used.

The non-bonded interactions between the atoms are described by Lennard-Jones potentials:

$$V = 4\varepsilon \left\{ \left(\frac{\sigma}{r}\right)^{12} - \left(\frac{\sigma}{r}\right)^{6} \right\} = \varepsilon \left\{ \left(\frac{r_m}{r}\right)^{12} - 2\left(\frac{r_m}{r}\right)^{6} \right\}. \tag{6.1}$$

The two different forms of the equation are related by $r_m = 2^{1/6}\sigma$, where σ is the first zero of the potential, while r_m is the position of the minimum (cf. Figure 6.2).

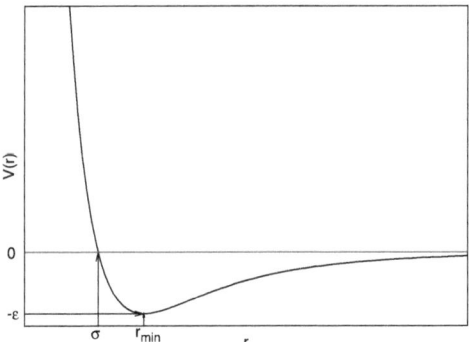

Figure 6.2.: Lennard-Jones potential.

In order to save computational time, we truncate the potential at a cut-off distance of 9 Å for the calculations of Lennard-Jones interactions.

We do not use a bond stretching potential. The integration time step is limited by the smallest oscillation period in the simulated system, *i.e.* the period of bond stretching oscillations. By constraining carbon-carbon bond lengths to their equilibrium length of 1.53 Å, we can remove the fastest degree of freedom and are able to use an integration time step of $dt = 2$fs without losing accuracy. The bond lengths are constrained using the LINCS algorithm [72].

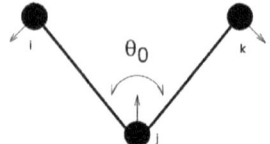

Figure 6.3.: Principle of angle bending.

The angle bending potential that is used in the force field is:

$$V_\alpha(\theta_{ijk}) = \frac{1}{2}k^\theta_{ijk}\left(cos(\theta_{ijk}) - cos(\theta^0_{ijk})\right)^2 \qquad (6.2)$$

where

$$cos(\theta_{ijk}) = \frac{\vec{b}_{ij} \cdot \vec{b}_{kj}}{|\vec{b}_{ij}||\vec{b}_{kj}|} \qquad (6.3)$$

and θ^0_{ijk} is the equilibrium value of the angle θ_{ijk}.

The dihedral angle ϕ of four consecutive atoms i, j, k, l is the angle between the plane containing the bonds $\vec{b}_{ij}$ and $\vec{b}_{jk}$, and the plane defined by the bonds $\vec{b}_{jk}$ and $\vec{b}_{kl}$. The principle is illustrated in Figure 6.4.

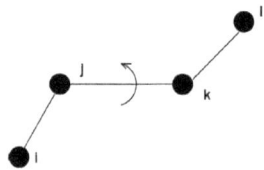

Figure 6.4.: Principle of dihedral angle.

For the dihedral interactions, a potential function based on expansion in powers of $cos(\phi)$ (Ryckaert-Bellemans type) is used.

$$V_{rb}(\phi_{ijkl}) = \sum_{n=0}^{5} C_n(cos(\psi))^n, \qquad (6.4)$$

6.1 MD Simulations of confined polyethylene melts

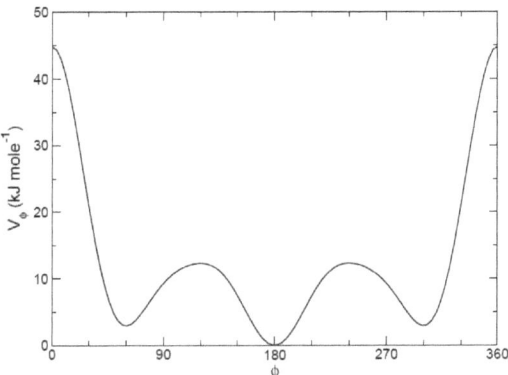

Figure 6.5.: Ryckaert-Bellemans potential.

where $\psi = \phi - 180°$. Reference [71] uses a different form of dihedral potential, so the parameters are translated accordingly to match the functional form of the potential used by GROMACS.
Simulations were carried out at a temperature of $T = 509$ K, using a Nosé-Hoover thermostat [73, 74].

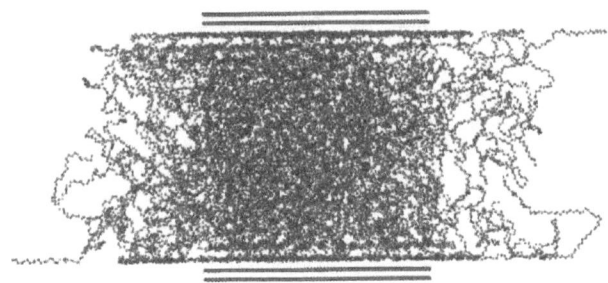

Figure 6.6.: Simulation snapshot of the confined polyethylene melt

6.1.2. Confined polyethylene chains

For the interaction between the graphite atoms and the polyethylene chains, graphite force field parameters were taken from reference [75], and the graphite-polyethylene Lennard-Jones interaction parameters were calculated using standard Lorentz-Berthelot mixing rules, *i.e.* an arithmetic average is used for length parameters, while a geometric average is used for energy parameters:

$$\sigma_{ij} = \frac{1}{2}(\sigma_i + \sigma_j) \tag{6.5}$$

$$\epsilon_{ij} = \sqrt{\epsilon_i \epsilon_j} \tag{6.6}$$

We set up the confined systems in the following way: As a first step, we carry out bulk molecular dynamics simulations of the melts at constant pressure ($p = 1$ atm) and constant temperature ($T = 509$ K). After equilibration, *i.e.* after the density and the gyration radius have reached constant values that are known from experiment and previous simulations studies [76, 77], we add two graphite double layers on each side in z-direction to the polyethylene melt. Two graphite layers are sufficient because a third layer would not be "visible" anymore by the monomers due to the cut-off for Lennard-Jones interactions.

Since for the bulk simulations, we use three-dimensional periodic boundary conditions, the whole chains extend over more than the size of the box in all three directions. For the confined system, we can use periodic boundary conditions in only two dimensions (x and y), whereas the third dimension z is finite and limited by the graphite surfaces. Thus, we can not simply place the surfaces in their target positions, because the chains would not fit in between the surfaces. We position the surfaces around the melt with an initial distance much higher than the target distance to avoid bad contact between monomers and graphite atoms. The distance is then reduced in very small steps ($\delta z = 0.1$ Å), and in between those steps, the system is equilibrated for 2 ps to adapt to the new constraints. The final wall distance is approximately a multiple of the radius of gyration ($d \approx 3 R_g$). The exact value is given by the final volume which is to correspond to the bulk volume of the system at $p = 1$ atm and the fact that the graphite layers have to fit exactly in the box to allow periodic boundary conditions.

The atoms that constitute the graphite double layers are restrained to their equilibrium positions during the simulations.

C100

First, we study a melt of $C_{100}H_{202}$, consisting of 40 chains confined between two parallel graphite double layers with distance $d \approx 3R_g^{C100} \approx 50\,\text{nm}$. The volume is kept constant to $V = 125\,\text{nm}^3$.

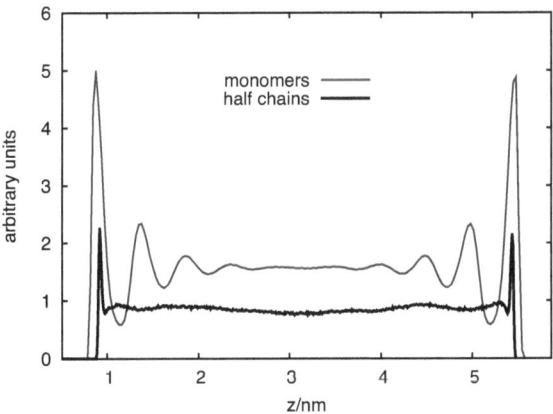

Figure 6.7.: Density profiles for half chains and monomers (C100).

We found a noticeable adsorption behavior when we investigated the half chain distribution of the C100 system: the half-chain distribution in Figure 6.7 shows narrow peaks close to the graphite surfaces, indicating that a small, but significant part of the half-chains is adsorbed in its complete length at one of the surfaces. If we compare the half-chain profile with the monomer profile shown in Figure 6.7, we see that these adsorbed chain parts are located completely in the first monomer layer.

The characteristic oscillatory behavior of the monomer density near the graphite walls has also been reported by previous studies [78, 79].

These results indicate that for such short chains, the energy gain close to the surfaces can outweigh the entropy loss by adsorption. Figure 6.8 shows a partial simulation snapshot of two C100 chains, each of them having one half adsorbed to a surface. This adsorption behavior made the C100 system unsuitable for our purpose, because a half-chain adsorbed at a surface is definitely not Gaussian in its behavior. Consequently, we decided to focus on longer chains with 300 backbone carbon atoms.

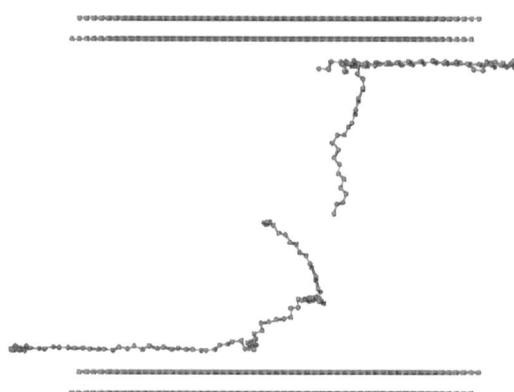

Figure 6.8.: Simulation snapshot of two C100 chains, each with one half adsorbed at the surface. The other chains in the system are stripped away for better visibility.

Admittedly, one could infer from the results above that for C300 there will still be chains with one third of a half-chain adsorbed at a surface, also leading to configurations far from being Gaussian. However, according to Daoulas et al. [79] who studied the conformations of polyethylene films adsorbed on graphite for various chain lengths by atomistic MC simulations, the shorter the chains, the longer connected chain parts will be fully adsorbed.
As we will see in the next paragraph, the time that the equilibration of C300 takes is already very long, making simulations of even longer chains hard to handle. C300 is hence the best trade-off between simulation time and expected results.

C300

We study a melt of 64 $C_{300}H_{602}$ chains confined between two graphite double layers with a distance $d \approx 3R_g \approx 90\,\text{nm}$. The volume is kept constant to $V = 581\,\text{nm}^3$.
The equilibration in this system took a very long time, so we decided to use the results we gathered after a simulated time of about $t \approx 1.5\,\upmu\text{s}$. The density profile of the whole-chain centers of mass shown in Figure 6.10 reveals the incomplete equilibration conditions. Figure 6.9 shows the probability profile to find the center of mass of a half chain in a given interval with distance z from the upper surface.

6.2 Derivation of the effective potential

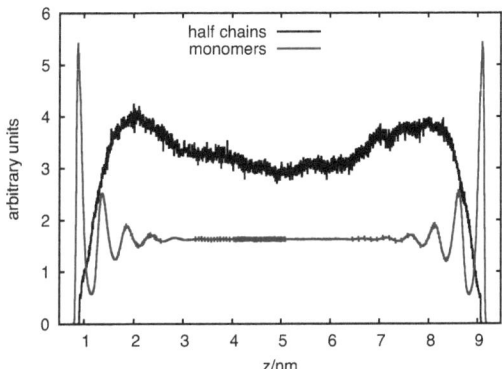

Figure 6.9.: Probability distribution profile of centers of mass of C300 half chains, distribution averaged over samples from the last 400 ns, and monomer density profile.

6.2. Derivation of the effective potential

The following section will describe how we derived an effective tabulated wall potential for the soft tetramer model, *i.e.* a potential that is able to reproduce the form of the distribution resulting from the detailed molecular simulations, based on the results from the previous section.

6.2.1. Iterative Boltzmann Inversion

Iterative Boltzmann Inversion is a method of systematic coarse-graining, using iterative inversion of radial distribution functions from experiments or atomistic simulations to derive effective coarse-grained potentials. It is based on a proposition by Soper [80] and has been further developed by Reith et al. [81].

The basic idea of Boltzmann inversion is the following: Assuming a given probability distribution obeys Boltzmann statistics, the probability $p(q) \propto \exp(-U(q)/k_B T)$. Thus, one can derive the potential of mean force $U(q)$ by taking the logarithm of the distribution:

$$U(q) = -k_B T \ln(p(q)) \qquad (6.7)$$

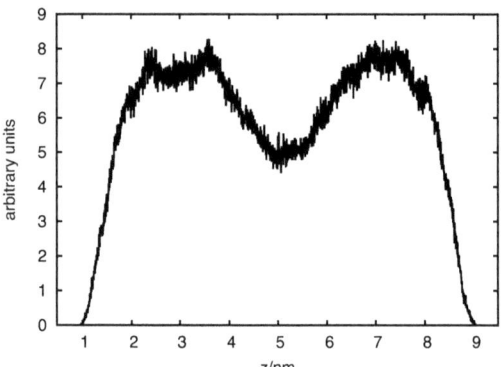

Figure 6.10.: C300 whole-chain center of mass density profile.

The potential of mean force is a free energy and not a potential energy, except for the – usually uninteresting – limit of infinite dilution, however it is sufficient as a initial guess $U_0(q)$ for an iterative procedure.

Similar to the method mentioned above, our approach uses a one-dimensional distribution of the simulated polymer melt between graphite walls as an input to derive an effective wall potential for the existing soft tetramer model that has been described in the previous chapters of this work.

6.2.2. Iteration procedure

In this section we describe the procedure for the derivation of an attractive interaction potential between the wall and the soft spheres, based on the distribution of half-chains that resulted from the detailed molecular simulations.

In our case, the target distribution function $g(z)$ is based on the distribution of C300 half-chain centers of mass shown in Figure 6.9. We symmetrize the distribution function by adding the content of bin n to the content of bin $(N - n)$ and dividing the sum by two (for all bins with a non-zero content). The symmetrized distribution is then smoothed by a five-point running average: every value is replaced by the arithmetic average of itself and the values of the four neighboring bins.

6.2 Derivation of the effective potential

Now, the distribution has to be mapped to the length scale of our soft-tetramer model. To this end, we use the relation

$$\sigma_{\alpha\alpha} = 2R_{g,\alpha} \qquad (6.8)$$

for $\alpha = \{A, B\}$ from chapter 3, Equation (3.15).

The surfaces in the atomistic simulation are separated by a distance of $d = 90\,\text{Å}$ which corresponds approximately to $3\,R_g^{C300}$. Because we look at parts of the C300 chains that correspond to only one half of the whole chain length, we have to compare the model length unit with the radius of gyration of a chain with 150 backbone carbon atoms. According to reference [77], $R_g^{C150} \approx 20\,\text{Å}$, hence $d \approx 4.5\,R_g^{C150}$ which corresponds to a wall distance of $\tilde{d} \approx 2.25\,\sigma$ in the soft-tetramer model, according to Equation (6.8).

Hence, by dividing the z value in angstroms of the symmetrized and smoothed C300 half chain distribution by $90/2.25 = 40$, we obtain the corresponding distribution in the length unit of our model, shown in Figure 6.11.

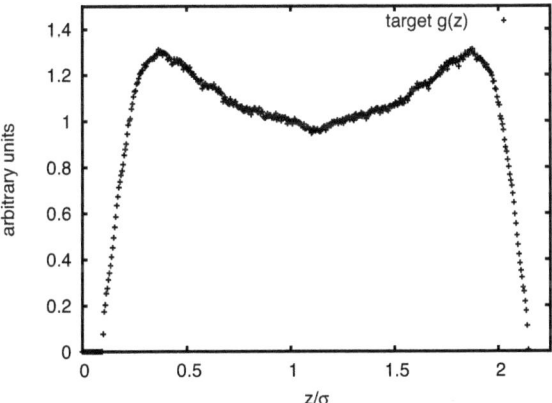

Figure 6.11.: Distribution from MD simulation (Figure 6.9), symmetrized, mapped to the soft-tetramer model length scale and smoothed by a five-point running average.

This is the target distribution function that an effective interaction potential between the wall and the soft spheres should yield when we use it for simulations of the confined soft-tetramer model.

We take the Boltzmann inverse of the distribution $g(z)$ shown in Figure 6.9 and use the result as an initial potential for the iterative procedure:

$$U_0(z) = -\ln(g(z))$$

Then we simulate a soft-tetramer melt confined between two walls with U_0 as a wall potential, which yields a distribution $g_0(z)$ that is different from $g(z)$. Figure 6.12 compares the distributions. If we compare the simulated distribution g_0 with the target distribution g, we see that the former rises too strongly for small distances from the wall, indicating that the initial potential for small distances from the wall is too attractive, whereas for larger distances g_0 is too low, meaning that U_0 is too repulsive in this part. We improve the potential for the next iteration step by forming the ratio of g_0 and g and then Boltzmann inverting it:

$$\delta U_1(z) = U_0(z) + \ln\left(\frac{g_0(z)}{g(z)}\right) \tag{6.9}$$

This correction term (also shown in Figure 6.12) is added to the test potential U_0.

We then iterate this step, using the following generalized iteration rule:

$$U_{i+1}(z) = U_i(z) + \delta U_{i+1}(z) = U_i(z) + \ln\left(\frac{g_i(z)}{g(z)}\right) \tag{6.10}$$

If the denominator is zero, U_{i+1} is set to a finite yet very high value to avoid that the algorithm crashes.

When the algorithm converges, we have a valid solution, meaning that the resulting U is an effective potential for the soft-tetramer model that is able to reproduce the form of the distribution from the detailed simulations.

We derive the effective wall potential for one temperature ($T = 1$), but use the same potential also for simulations at somewhat lower temperatures. This approximation is valid as long as the entropic part of the interaction does not depend too strongly on temperature. Within the weak segregation regime where we want to perform our simulations, this assumption can be considered as reasonable.

For the iteration procedure we use a symmetric soft tetramer melt with $f = 0.5$ and density $\rho_c = 1$, that is one tetramer chain per unit volume, in the disordered state at $T = 1$, meaning $U_{AA} = U_{BB} = U_{AB}$, between two walls in z direction with distance $L_z = 2.25\sigma$, with the same wall interaction for both blocks.

The z coordinate is divided into intervals of size 0.005σ. A five-point running average is

6.2 Derivation of the effective potential

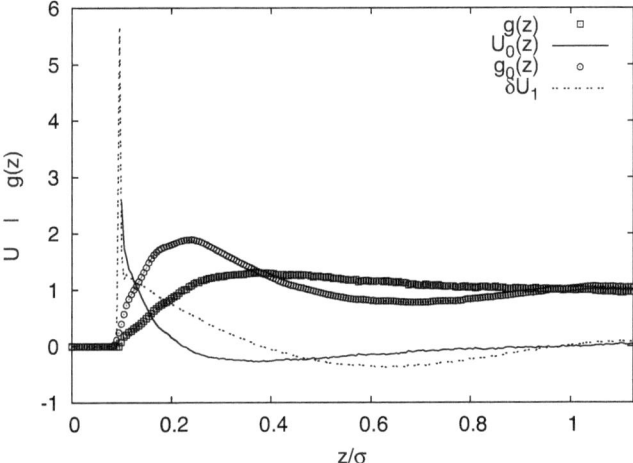

Figure 6.12.: First iteration: the target distribution g_0 is represented by open squares, its Boltzmann inverse, the initial potential U_0, by a solid line. The resulting distribution from the first iteration, g_1, is represented by open circles. The dashed line δU_1 has to be added to U_0 in order to improve potential and get the wall potential U_1 for the next iteration.

used in every iteration step to smooth the simulated distributions.

6.2.3. Root mean square deviation

To measure the difference between the simulated distribution and the target distribution, in every iteration the root mean square deviation (RMSD) is calculated:

$$RMSD(i) = \sqrt{\frac{\sum_{z}^{n}(g(z) - g_i(z))^2}{n}} \qquad (6.11)$$

Since it turns out that the algorithm tends to overestimate the correction term $\delta U_i(z)$, so that the RMSD increases again after an initial decrease, a correction factor w has been

introduced:
$$U_{i+1}(z) = U_i(z) + w \cdot \delta U_{i+1}(z) \qquad (6.12)$$

As a first guess, w was chosen to be 0.3, which turned out to be a fairly suitable value as a trade-off between computing time and resulting agreement that can be reached.

Figure 6.13 shows the root mean square deviation vs. iteration. After a strong initial decline it converges quickly and fluctuates around 0.05. Since this is close enough for our purpose, we take the potential from the iteration at which the RMSD seems to reach the plateau and stops decreasing significantly, which we determine to be iteration 21.

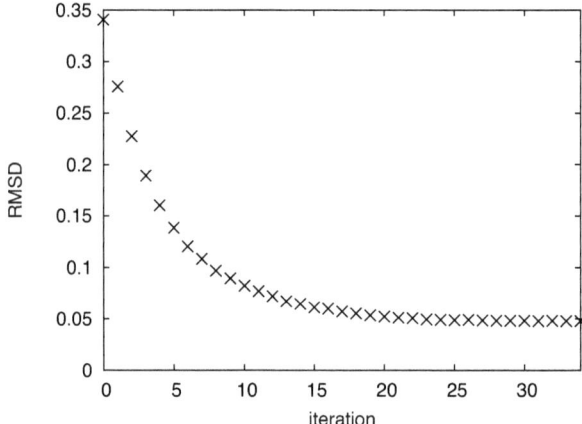

Figure 6.13.: The root mean square deviation reaches a plateau at iteration 21.

6.2.4. Effective potential

To be able to employ different wall potentials for both blocks, we proceed in the following way: We take the potential from iteration 21 as the attractive potential for one of the blocks, after smoothing it again with a five point running average. For the other block we use a purely repulsive potential that we obtain by truncating the smoothed attractive

6.2 Derivation of the effective potential

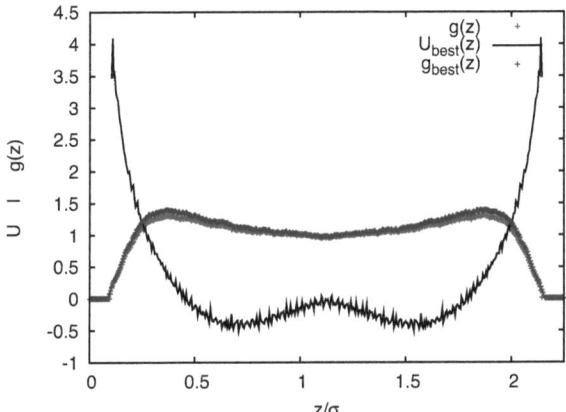

Figure 6.14.: Final iteration: After 21 iterations, the distribution $g_{\text{best}}(z)$ that has been generated using $U_{\text{best}}(z) = U_{21}(z)$ (solid line) resembles the target distribution $g(z)$ so that it is almost impossible to tell them apart.

potential at its minimum and shifting it by the depth of the potential well $\epsilon \approx 0.5$ (see Figure 6.15).

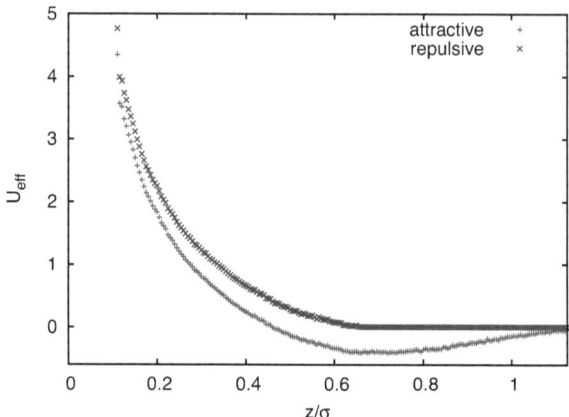

Figure 6.15.: Effective potential used for attractive interaction ($r_{\min} \approx 0.7, \epsilon \approx 0.5$) and the same potential shifted and cut off at the minimum for purely repulsive interaction.

6.3. Soft-tetramer model in confinement

In this section, we present results for simulation studies of symmetric diblock copolymer thin films confined between two surfaces, using the soft tetramer model. We investigate a compositionally symmetric system ($f = 0.5$) at $\chi N = 24$ and $\rho_c = 1$. Two kinds of surfaces are considered: symmetric surfaces preferential to A blocks and purely repulsive to B blocks, and symmetric neutral surfaces repelling both blocks with the same surface–block interaction. We use two-dimensional periodic boundary conditions in x and y direction with the surfaces confining the simulation box in z direction.

The morphology of the confined diblock copolymer melt will be discussed in the following sections.

From the results of the bulk simulations described in chapter 5 at the same conditions that are used in the confined soft-tetramer model simulations (control parameters $f = 0.5$, $\chi N = 24$, $L = 12$), the bulk lamellar period, L_0, is determined using the cluster algorithm on discretized configurations. This method is described in section 5.3. The distribution of $< G_{3,AB}^{\mathrm{mapped}} >$ peaks at 0.29, as shown in Figure 6.16. According to Equation (4.3), this

6.3 Soft-tetramer model in confinement

corresponds to a lamella thickness of $L_d = 1.86$, hence the bulk lamellar period can be determined as $L_0 = 2 \cdot L_d = 3.72$.

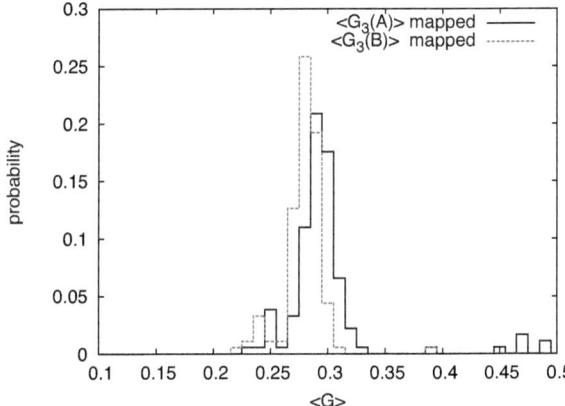

Figure 6.16.: Distribution of $\langle G_3^{AB} \rangle$ resulting from the application of the cluster algorithm to discretized configurations for the bulk system $f = 0.5$, $\chi N = 24$, $\rho_c = 1$.

6.3.1. Preferential walls

We investigate wall interactions that are preferential to one block and repulsive to the other block.

We vary the surface separation from $L_z = 4\sigma$ to $L_z = 14\sigma$ in steps of $0.5\sigma - 1\sigma$. For all surface separations, we find lamellae parallel to the walls as a stable morphology, with the preferred block at the wall, their number depending on L_z: the number of bilayers is an integer between $L_z/L_0 - 1/2$ and $L_z/L_0 + 1/2$.

The more L_z and L_0 are compatible with each other, the closer is the observed lamellar period to the bulk lamellar period. When L_z/L_0 is not an integer, the confined lamellar period deviates from L_0 to accommodate the frustration. The closer L_z/L_0 to half an odd integer, the stronger the frustration.

Figures 6.17 and 6.18 summarize the results by showing the profile along the z-axis of the probability to find an A or B sphere at a given z for all surface separations L_z. For the respective values of L_z/L_0, refer to Table 6.1

The profiles in Figures 6.17 and 6.18 show that the lamellae are compressed and stretched to fit between the surfaces for each surface distance, and that additional lamellae develop when the surface distance reaches certain values.

We observe an increased density close to the walls and a depletion at A-B interfaces that is visible in figures 6.17 and 6.18.

For the surface separations with the strongest frustration, the central lamellar layer becomes defective and interconnected. These values can be recognized in Figure 6.17 as $L_z = 5.5$, $L_z = 6$ and $L_z = 9.5$ and in Figure 6.18 as $L_z = 13$ because for these values, the distribution of one component does not vanish completely when the distribution of the other component peaks. Table 6.1 summarizes the results and shows that, for the separations leading to defective lamellae, L_z/L_0 is the closest to half an odd integer.

Figure 6.19 shows exemplary simulations snapshots for some selected surface distances. The behavior discussed above is visible, like the depletion at A-B interfaces, the cyclic compressing and stretching of the lamellae and the developing of new layers as a function of L_z, and the defective interconnected lamellae for frustrating surface distances.

In order to make sure that there are no other (meta-)stable states, we have carried out additional simulations starting from a configuration with lamellae perpendicular to the walls, however the results that we obtained were exactly the same as with the usual isotropic start configuration. We conclude that the results described above are the only stable states.

6.3 Soft-tetramer model in confinement

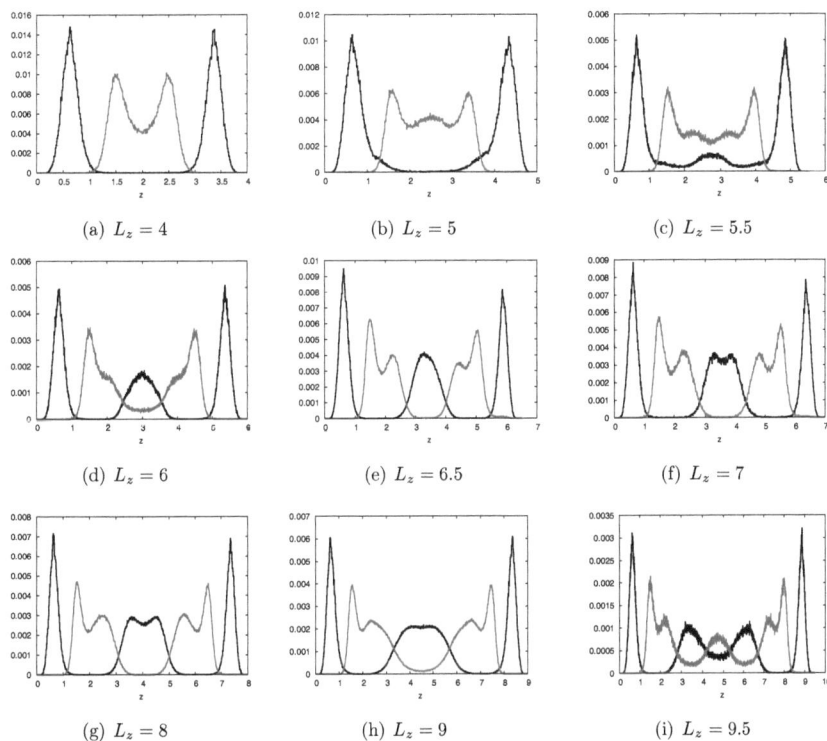

Figure 6.17.: A (black) and B (gray) sphere distribution profiles for various surface distances L_z. For surface distances $L_z = 5.5$, $L_z = 6$ and $L_z = 9.5$, frustration leads to defective lamellae. The x-axis corresponds to the z-coordinate and the y-axis to the probability of finding the center of an A or B sphere at a given z.

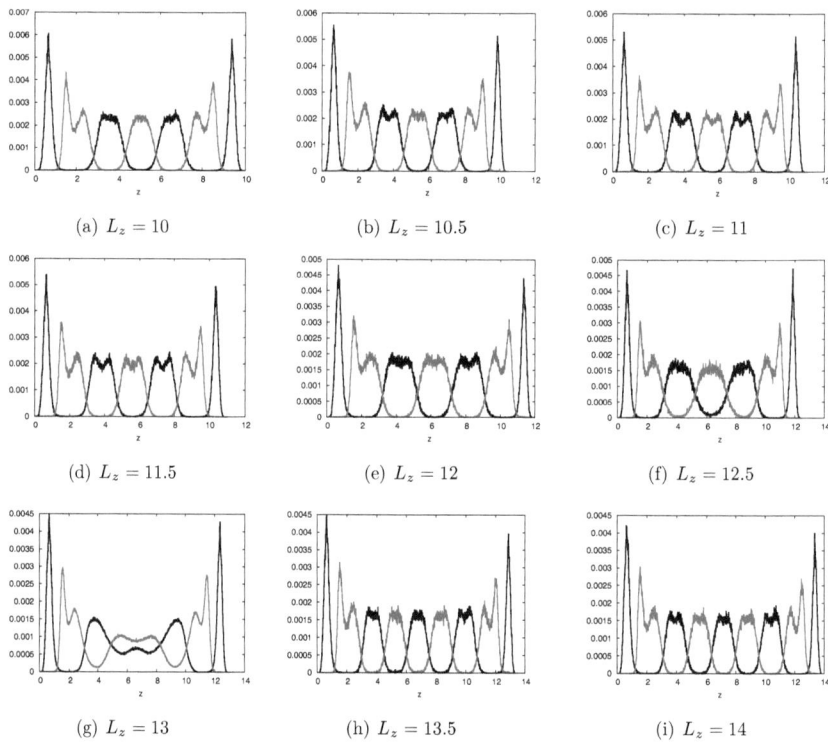

Figure 6.18.: A (black) and B (gray) sphere distribution profiles for various surface distances L_z. For surface distance $L_z = 13$, frustration leads to defective lamellae. The x-axis corresponds to the z-coordinate and the y-axis to the probability of finding the center of an A or B sphere at a given z.

6.3 Soft-tetramer model in confinement

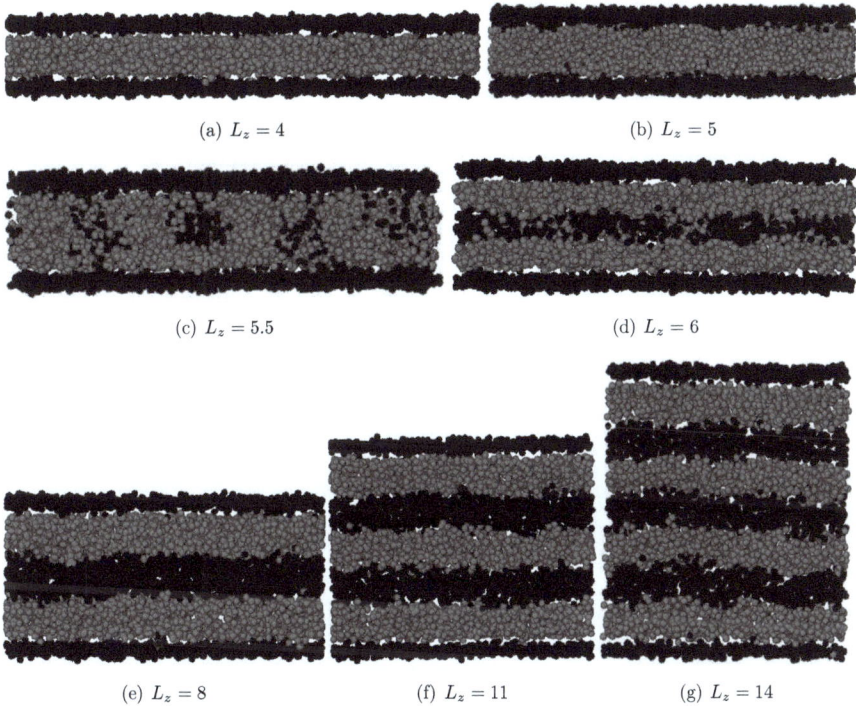

Figure 6.19.: Simulation snapshots for selected surface distances L_z. In each line, figures are to scale. Figures (c) and (d) show frustrated defective lamellae in the central layer.

L_z	L_z/L_0	number of bilayers
4	1.08	1
5	1.34	1
5.5	1.48	defective
6	1.61	defective
6.5	1.75	2
7	1.88	2
8	2.15	2
9	2.42	2
9.5	2.55	defective
10	2.69	3
10.5	2.82	3
11	2.96	3
11.5	3.09	3
12	3.23	3
12.5	3.36	3
13	3.49	defective
13.5	3.63	4
14	3.76	4

Table 6.1.: Results for the $f = 0.5, \chi N = 24, \rho_c = 1$ system confined between preferential surfaces. Frustration is strongest the closer L_z/L_0 to half an odd integer, inducing defective layers. The number of bilayers is an integer between $L_z/L_0 - 1/2$ and $L_z/L_0 + 1/2$.

6.3.2. Neutral walls

We simulate the same system with neutral walls (purely repulsive to both blocks) for the same surface separations as above, and we find lamellae perpendicular to the walls for all distances.

An example simulation snapshot for surface separation $L_z = 7$ is shown in Figure 6.20(a). Figure 6.20(b) shows a distribution profile for the same surface separation. There is still an increase in density near the surface due to the hard walls, however less pronounced than in the case of the preferential surfaces.

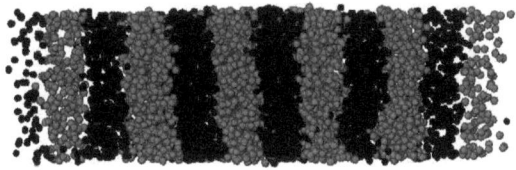

(a) Example simulation snapshot

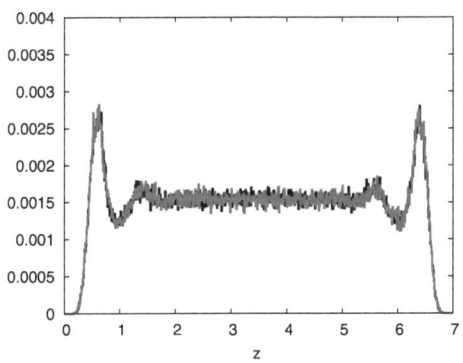

(b) A (black) and B (gray) sphere distribution profile

Figure 6.20.: Confinement of the soft-tetramer model between neutral walls results in lamellae perpendicular to the walls, $L_z = 7$.

6.3.3. Discussion

The results we find for neutral surfaces show good agreement with experiment, theory and previous simulation studies.

Wang et al. [82] carried out Monte Carlo simulations for a simple cubic lattice model, and they also found perpendicular lamellae for all surface distances for a diblock melt confined between two neutral surfaces.

Comparable results have also been found in experiments [83], where the elimination of preferential interactions at both the substrate and surface interfaces leads to a perpendicular orientation of the microdomains persisting through the entire film.

Theoretical studies using self-consistent field methods also predict perpendicular lamellae to be the most stable morphology regardless of the surface separation [84, 85]. It has been argued that this is due to nematic ordering of the monomers arising from the orientational constraint imposed by the walls [84].

For the preferential surfaces on the other hand, the agreement is less clear. All references to which we have compared our results agree in the fact that there is a close connection between the amount of frustration due to incommensurate surface distances and the observed structure of the confined diblock melt. There is however less accordance on how exactly the structure is affected in cases where the frustration is strong.

Compared with the results for the simple cubic lattice model by Wang et al. [82], our results for preferential surfaces are very similar for weak frustration, but where we find that strong frustration leads to defective and interconnected lamellar layers, they report lamellae perpendicular to the walls.

Other lattice Monte Carlo studies [86, 87] find that the lamellar order will be suppressed or deformed due to the conflict between the two lengths and result in tilted *or* deformed lamellar structure, or even coexistence of lamellae in different orientations are found in cases of strong conflict.

Theory predicts perpendicular lamellae for frustrating distances, but with distorted A-B interfaces [84].

Experiments [88] have also reported perpendicular lamellae for weak surface interactions, similar to those found by reference [82]. For strong interaction however, they saw partially disordered and interconnected structures between preferential surfaces induced by frustration, which is comparable to what we find for the central lamellar layer.

6.3 Soft-tetramer model in confinement

Although there are some minor contradictions and it is unclear to which extent the agreement can be reached, altogether, our model is able to reproduce the general features for confined films of diblock copolymer melts that have been found by previous studies.

Within the scope of this work, we only have investigated the confinement behavior at $\chi N = 24$. It is possible that for lower values of χN, the results for preferential walls will look different, because the preferred component will be pushed to the wall to a lesser extent. A more thorough study that extends to a wider temperature range is left for future work.

CHAPTER 7

Conclusions

We have introduced a soft-tetramer model for diblock copolymer melts with effective interactions at the scale of the radius of gyration. This model is able to reproduce the form of the experimentally found diblock copolymer phase diagram.

We have studied our model at two different densities that lead to invariant degrees of polymerization N_{inv} for which strong deviations from the mean field phase diagram are to be expected. In contrast to the mean-field prediction (see Figure 2.3(b)), we have observed direct first-order transitions from disordered melt to lamellae, cylinder and perforated lamellae phases, much like those identified in experiments (see Figure 2.3(a)). We have compared our results with other studies in the same range of invariant degree of polymerization, and have found that our results are in good qualitative accordance. We conclude therefore that our model can capture the fluctuations that are responsible for the differences between the phase diagrams from self-consistent field theory and from experiments, meaning that these are fluctuations on the scale of the radius of gyration.

While previous molecular simulation studies have been limited to the investigation of a narrow parameter regime, with our simple and efficient model we have been able to determine the equilibrium morphology of the diblock copolymer melt at a wide range of phase points in the $(f, \chi N)$ space, which also allows us to study phase transitions not only from the disordered into the ordered phase, but also between the observed ordered phases.

Beyond its ability to correctly reproduce the features of the experimental diblock copolymer phase diagram, our model is able to measure the changes of intramolecular structure, namely a gradual stretching of the chains in the disordered phase upon approaching the disorder-order transition point, which is an indication of the deficiency of the random phase approximation for the description of the diblock copolymer phase diagram. The observed effect is of the same size as it has been seen in previous molecular simulations.

The mesophases have been clearly identified using a novel approach that formalizes the method of direct geometric observation which is used commonly in experiments and molecular simulation studies. The visual impression of a simulation snapshot is described by the distribution of the number and shape of clusters formed by the components of the diblock. We have shown that this method allows us to clearly determine the mesophase morphology. It has been suggested to use the Euler characteristic for the purpose of mesophase identification, however we have shown that its discriminating power is lower than that of our cluster analysis method.
Although the Euler characteristic proved to be less beneficial than expected, we could use the discretization method that we have developed for its calculation for other analyses that require discretized configurations.

After deriving an effective interaction potential acting between the soft spheres and a surface from the results of atomistic simulations, an initial study has shown that the model can also be used for the simulation of confined diblock copolymer melts, yielding results that are comparable to what has been found in previous experimental, theoretical and simulation studies. Within the scope of this work, we investigated the confined soft-tetramer model only for a single value of χN, although it might be interesting to study the confinement behavior at a broader range of temperatures.

Besides the last mentioned issue, this work offers several interesting possibilities to continue research in various directions:
By construction, the model will be well suited for the investigation of structure formation in composite systems consisting of nano-sized filler particles dissolved in a block copolymer

matrix. Nanoparticles and diblock chains will be modeled on the same length scale, which permits very efficient simulations.

In the scope of this work, we have investigated the effect of parallel confining graphite surfaces on the distribution of polyethylene half chains, and have derived an effective potential that describes the interaction between parallel flat surfaces and the spheres of the soft-tetramer model. This can be done in a similar way for other geometries, for example with curved surfaces. In this way, the effective interactions between a nanoparticle and a soft tetramer can be derived. As an alternative, one can try to deduce the form of the potential for the curved geometry based on the potential for the flat surfaces.

Our approach, however, will only work if the nanoparticles stay well separated in the block copolymer matrix. If surfaces come closer than the radius of gyration, the system can not be described by means of this model.

We have not been able to observe the gyroid phase. In the relevant region of the control parameter space we only found the perforated lamellae phase. Since we believe that this is due to the incommensurate box dimensions, it would be interesting to gradually vary the size of the simulation box until it fits to the gyroid lattice spacing, and compare the free energies of both phases.

Increasing the chain density up to values of about $\rho_c \approx 7$ would lead to an invariant degree of polymerization for which fluctuations become less relevant, so it may be interesting, albeit computationally expensive, to carry out simulations at that range of densities. Finally, in order to further prove the validity of the model, it may be useful to perform mean field calculations for the soft-tetramer model, and to compare the results to the results of the simulations.

APPENDIX A

Cluster analysis for disordered states

Although we can not know a priori the number and shape of the clusters in the disordered phase, we can use empirical knowledge to identify the disordered phase. In the compo-

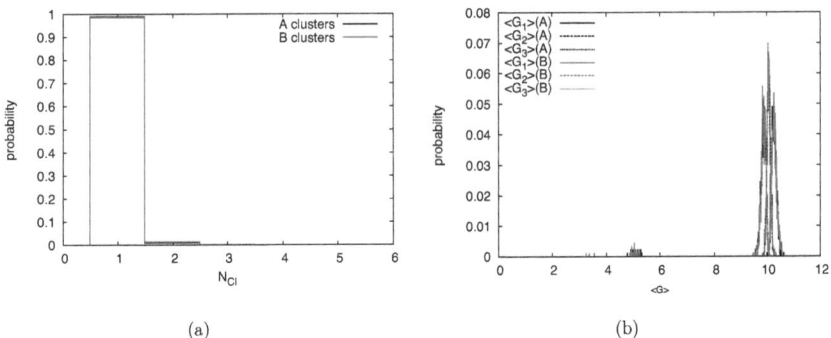

Figure A.1.: Distributions of cluster numbers (a) and gyration tensor eigenvalues (b) in the disordered phase. $f = 0.5$, $\chi N = 1$, $\rho_c = 1.3$, $L = 11$.

sitionally symmetric and high temperature case ($f = 0.5$, $\chi N = 1$), the configuration is basically isotropic, which can also be seen in the snapshot in Figure A.2. There exists only one cluster for each component (Figure A.1(a)), extending over all of the simulation box (Figure A.1(b)).

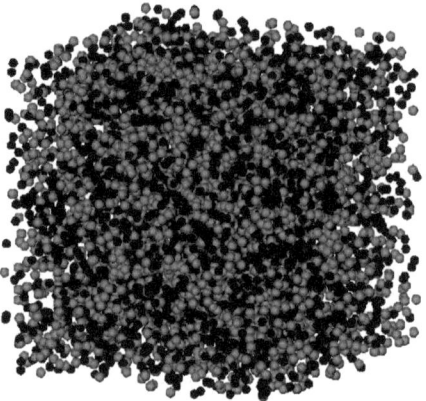

Figure A.2.: Simulation snapshot at parameters $f = 0.5$, $\chi N = 1$, $\rho_c = 1.3$, $L = 11$.

Upon approaching the disorder-order transition however, as we show for $f = 0.5$, $\chi N = 6$, the clusters start to fall apart in two or even three clusters in some of the sample configurations (Figure A.3(a)), and for these samples, we obtain smaller mean eigenvalues and get three distinct peaks in the mean eigenvalue distribution (Figure A.3(b)).

Due to the fact that the next neighbor radius for the cluster analysis algorithm is different for A and B clusters (cf. chapter 4.1.1.2), the situation looks quite different for higher compositional asymmetries f. As an example, we show the distributions for $f = 0.81$, $\chi N = 21$ in Figure A.4. The cluster number distributions for in Figure A.4(a) look vaguely similar to what we saw for cylinders, however much broader, and if we additionally look at the mean eigenvalue distributions in Figure A.4(b), there is no risk of confounding the disordered phase with the cylinder phase, because all of the eigenvalues of the minority clusters $\langle G_{i,B} \rangle$ basically amount to the same value, and the characteristic behaviour of the cylinder phase with one significantly higher and two small mean eigenvalues is missing. Averaging over arbitrarily shaped clusters yields mean eigenvalues $\langle G_1 \rangle \approx \langle G_2 \rangle \approx \langle G_3 \rangle$ with a broad spread for each eigenvalue. A simulation snapshot is shown in Figure A.5.

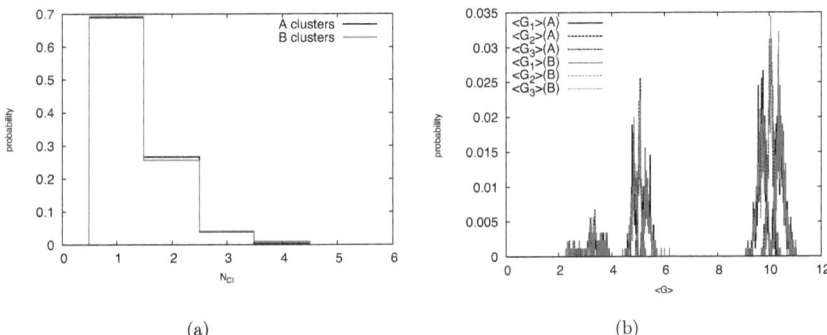

Figure A.3.: Distributions of cluster numbers and gyration tensor eigenvalues in the disordered phase. $f = 0.5$, $\chi N = 6$, $\rho_c = 1.3$, $L = 11$.

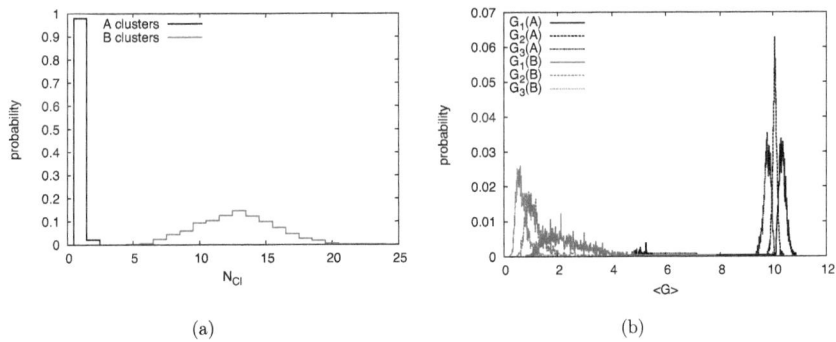

Figure A.4.: Distributions of cluster numbers and gyration tensor eigenvalues in the disordered phase. $f = 0.81$, $\chi N = 21$, $\rho_c = 1.3$, $L = 11$.

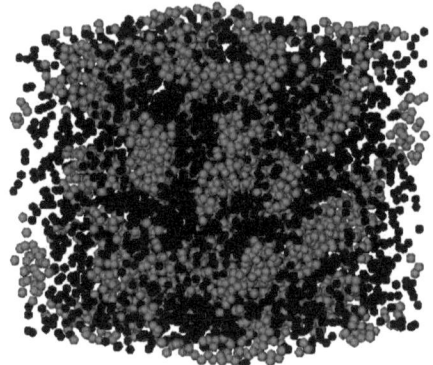

Figure A.5.: Simulation snapshot at parameters $f = 0.81$, $\chi N = 21$, $\rho_c = 1.3$, $L = 11$.

APPENDIX B

Grand-canonical Monte Carlo moves

When we try to insert a randomly grown chain into a rather dense system of chains, the probability is high that the move will be rejected. To overcome this problem, one employs a method that lets the chain grow into directions that are energetically favorable. However, this introduces an unphysical bias that has to be taken care of by determining the correct Monte Carlo acceptance rule.

B.1. Simulation technique

We choose with equal probability to insert or delete a tetramer chain. We follow the respective algorithms for insertion and deletion:

Insertion

- We insert the first sphere of the chain at a position chosen at random.

- We choose a bond length randomly with a probability according to the Boltzmann distribution, then generate $k = 10$ trial vectors $\{\vec{b}_i\}$ with a length equal to the chosen bond length evenly distributed around the first sphere, so that we end up with k possible positions for the next sphere.

- One of the k trial positions is chosen according to its Boltzmann weight, $w_i = e^{-\beta U(\vec{b}_i)}$.
- We repeat the last two steps until the new tetramer chain is complete.
- The new tetramer chain is accepted with acceptance rate $A_{N \to N+1}$.

Deletion

- One of the chains existing in the system is selected at random.
- We generate $k-1$ evenly distributed trial vectors with length equal to the bond length of the last bond, around the last sphere in the chain. Including the true bond, we have k vectors $\{\vec{b}_i\}$ which will be needed later for the determination of the acceptance rule.
- The last sphere is deleted from the chain.
- We repeat the last two steps until we reach the first sphere in the tetramer chain.
- The deletion of the chain is accepted with acceptance rate $A_{N+1 \to N}$.

Acceptance rate

To determine the acceptance rates, we use the condition of detailed balance (*cf.* chapter 5.1.1), which in this case has the extended form:

$$P(N) P_{\text{gen}}(N \to N+1) A_{N \to N+1} = P(N+1) P_{\text{gen}}(N+1 \to N) A_{N+1 \to N} \quad \text{(B.1)}$$

where P_{gen} is the probability of performing a certain trial move. If we write

$$\frac{A_{N \to N+1}}{A_{N+1 \to N}} = \frac{P(N+1)}{P(N)} \cdot \frac{P_{\text{gen}}(N+1 \to N)}{P_{\text{gen}}(N \to N+1)} =: y, \quad \text{(B.2)}$$

the acceptance rules

$$A_{N \to N+1} = \min(1, y), \quad \text{(B.3)}$$

$$A_{N+1 \to N} = \min(1, 1/y) \quad \text{(B.4)}$$

will obey the condition of detailed balance in Equation (B.2).

B.1 Simulation technique

The grand-canonical probabilities of a system with N chains and a system with $N+1$ chains relate as follows:

$$\frac{P(N+1)}{P(N)} = e^{\beta\mu} e^{-\beta U(\text{chain})}, \tag{B.5}$$

and the probability for the generation of a move that tries to delete one given chain out of $N+1$ is

$$P_{\text{gen}}(N+1 \to N) = \frac{1}{2}\frac{1}{N+1}, \tag{B.6}$$

where the factor $1/2$ expresses the 50% probability of a deletion move as opposed to an insertion move.

The probability for the generation of a move that tries to insert a chain is given by

$$P_{\text{gen}}(N \to N+1) = \frac{1}{2}\frac{1}{V}\prod_{n=1}^{3}\frac{ke^{-\beta U_{\text{bond}}(\vec{b}_n)}}{\int d\vec{b}_n e^{-\beta U_{\text{bond}}(\vec{b}_n)}}\frac{e^{-\beta U_{\text{nb}}(\vec{b}_n^{in})}}{\sum_{j=1}^{k} e^{-\beta U_{\text{nb}}(\vec{b}_n^j)}}, \tag{B.7}$$

where $U_{\text{nb}}(\vec{b}_n)$ signifies the energy of the sphere s_{n+1} following bond $\vec{b}_n$. The factor $1/2$, as in the deletion case, expresses the 50% probability of choosing an insertion move over a deletion move, the factor $1/V$ accounts for the first sphere which can placed anywhere into the simulation box with volume V. The product runs over the three bonds in the tetramer chain and describes the probability of choosing the k trial vectors times the probability to select the vector $\vec{b}_n^{in}$ out of them for each of the three bonds.

Expanding the fraction in Equation (B.7) with the Boltzmann weight of the first sphere, $e^{-\beta U_{nb}(s_1)}$, and introducing the Rosenbluth weight of the whole new chain

$$R = e^{-\beta U_{nb}(s_1)}\prod_{n=1}^{3}\left(\frac{1}{k}\sum_{j=1}^{k} e^{-\beta U_{\text{nb}}(\vec{b}_n^j)}\right), \tag{B.8}$$

we can reduce Equation (B.7) to

$$P_{\text{gen}}(N \to N+1) = \frac{1}{2V}\frac{e^{-\beta U(\text{chain})}}{Z_{\text{bond}}^3}\frac{1}{R}. \tag{B.9}$$

Combining Equations (B.2), (B.5), (B.6) and (B.9) and replacing $\mu' = \mu + \frac{3}{\beta}\ln(Z_{\text{bond}})$ we obtain the expression for y that we have been searching:

$$y = \frac{V}{N+1} R e^{\beta\mu'} \tag{B.10}$$

By substituting Equation (B.10) in Equations (B.3) and (B.4), we finally obtain the expressions for the acceptance rates:

$$A_{N \to N+1} = \min(1, \frac{V}{N+1} R e^{\beta \mu'}), \tag{B.11}$$

$$A_{N+1 \to N} = \min(1, \frac{N+1}{RV} e^{-\beta \mu'}). \tag{B.12}$$

APPENDIX C

The molecular dynamics method

In this chapter, we outline briefly the principles of the molecular dynamics simulations that have been used for the part of this work described in chapter 6.1. A thorough description can be found in the GROMACS manual [89].

Molecular dynamics simulations consist in numerically solving Newton's equation of motion for a system of N interacting particles:

$$m_i \frac{d^2 \vec{r}_i}{dt^2} = \vec{F}_i \qquad \text{(C.1)}$$

with $i = 1, \ldots N$, and the forces $\vec{F}_i$ are the negative derivatives of a potential function $U = U(\vec{r}_1, \ldots \vec{r}_N)$:

$$\vec{F}_i = -\nabla U. \qquad \text{(C.2)}$$

This potential function is a combination of all the relevant interactions given in chapter 6.1. The equations are solved simultaneously in small time steps δt. The system is followed for some time and the positions and momenta of the particles are written down at regular intervals. All observables that we want to measure have to be expressed as a function of these positions and momenta.

After all forces at a certain time t have been calculated, the positions and velocities need to be updated. This is done in GROMACS using the so-called "leap-frog" integration algorithm [90] that uses particle positions and forces determined by them at time t and velocities at time $t - \delta t/2$:

$$v(t + \frac{\delta t}{2}) = v(t - \frac{\delta t}{2}) + \frac{\vec{F}(t)}{m} \delta t, \qquad (C.3)$$

$$r(t + \delta t) = r(t) + v(t + \frac{\delta t}{2}) \delta t. \qquad (C.4)$$

To enable canonical simulations, *i.e.* to maintain constant temperature within the system, the equations of motion are modified for temperature coupling, in our case the Nosé-Hoover temperature coupling scheme [73, 74] that is supported by GROMACS. The equations of motion are extended by a friction term that is proportional to the particle's velocity and a "heat bath variable" ξ:

$$\frac{d^2 \vec{r}_i}{dt^2} = \frac{\vec{F}_i}{m} - \xi \frac{d\vec{r}_i}{dt}. \qquad (C.5)$$

The variable ξ is described by an equation of motion of its own; its time derivative is calculated from the difference between the reference temperature T_0 and the current temperature of the system T:

$$\frac{d\xi}{dt} = \frac{1}{Q}(T - T_0). \qquad (C.6)$$

The temperature of the system is proportional to its total kinetic energy:

$$N_{df} \frac{k_B T}{2} = E_{kin} = \frac{1}{2} \sum_{i=1}^{N} m_i v_i^2 \qquad (C.7)$$

where N_{df} is the number of degrees of freedom which can be calculated as $N_{df} = 3N - N_c$, with N_c the number of constraints imposed on the system. The constant Q determines the strength of the coupling. The Nosé-Hoover temperature coupling approach leads to an oscillatory relaxation. The period τ of the oscillations of kinetic energy between the system and the thermal reservoir is given by

$$Q = \frac{\tau^2 T_0}{4\pi^2}. \qquad (C.8)$$

Bibliography

[1] R. A. Segalman, B. McCulloch, S. Kirmayer, and J. J. Urban. Block Copolymers for Organic Optoelectronics. *Macromolecules*, 42(23):9205–9216, Dec 8 2009.

[2] M. R. Bockstaller, R. A. Mickiewicz, and E. L. Thomas. Block copolymer nanocomposites: Perspectives for tailored functional materials. *Advanced Materials*, 17(11):1331–1349, Jun 6 2005.

[3] Y. Lin, A. Boker, J. B. He, K. Sill, H. Q. Xiang, C. Abetz, X. F. Li, J. Wang, T. Emrick, S. Long, Q. Wang, A. Balazs, and T. P. Russell. Self-directed self-assembly of nanoparticle/copolymer mixtures. *Nature*, 434(7029):55–59, Mar 3 2005.

[4] B. de Boer, U. Stalmach, P. F. van Hutten, C. Melzer, V. V. Krasnikov, and G. Hadziioannou. Supramolecular self-assembly and opto-electronic properties of semiconducting block copolymers. *Polymer*, 42(21):9097–9109, Oct 2001.

[5] D. R. Sadoway. Block and graft copolymer, electrolytes for high-performance, solid-state, lithium batteries. *Journal of Power Sources*, 129(1):1–3, Apr 15 2004.

[6] N. P. Balsara and M. J. Park. Block copolymers and nanotechnology. *Journal of Polymer Science Part B-Polymer Physics*, 44(24):3429–3430, Dec 15 2006.

[7] P.-G. de Gennes. *Scaling concepts in polymer physics*. Cornell University Press, 1979.

[8] J. C. Le Guillou and J. Zinn-Justin. Critical Exponents for n-Vector Model in 3 Dimensions from Field-Theory. *Physical Review Letters*, 39(2):95–98, 1977.

[9] J. P. Wittmer, P. Beckrich, A. Johner, A. N. Semenov, S. P. Obukhov, H. Meyer, and J. Baschnagel. Why polymer chains in a melt are not random walks. *EPL*, 77(5), 2007.

[10] M. Doi. *Introduction to Polymer Physics*. Oxford University Press, Oxford, U.K., 1996.

[11] F. Schmid. Theory and simulation of multiphase polymer systems. In *Handbook of Multiphase Polymer Systems*. Wiley-Blackwell, Boston, 2010.

[12] G. S. Grest and K. Kremer. Molecular dynamics simulation for polymers in the presence of a heat bath. *Phys. Rev. A*, 33(5):3628–3631, May 1986.

[13] K. Kremer and G. S. Grest. Dynamics of Entangled Linear Polymer Melts - A Molecular-Dynamics Simulation. *Journal of Chemical Physics*, 92(8):5057–5086, Apr 15 1990.

[14] Q. Wang and Y. Yin. Fast off-lattice Monte Carlo simulations with "soft" repulsive potentials. *Journal of Chemical Physics*, 130(10), Mar 14 2009.

[15] M. Murat and K. Kremer. From many monomers to many polymers: Soft ellipsoid model for polymer melts and mixtures. *Journal of Chemical Physics*, 108(10):4340–4348, Mar 8 1998.

[16] F. Eurich and P. Maass. Soft ellipsoid model for Gaussian polymer chains. *Journal of Chemical Physics*, 114(17):7655–7668, May 1 2001.

[17] T. Vettorel, G. Besold, and K. Kremer. Fluctuating soft-sphere approach to coarse-graining of polymer models. *Soft Matter*, 6(10):2282–2292, 2010.

[18] C. I. Addison, J. P. Hansen, V. Krakoviack, and A. A. Louis. Coarse-graining diblock copolymer solutions: a macromolecular version of the Widom-Rowlinson model. *Molecular Physics*, 103(21-23):3045–3054, Nov-Dec 2005.

[19] C. Pierleoni, C. Addison, J. P. Hansen, and V. Krakoviack. Multiscale coarse graining of diblock copolymer self-assembly: From monomers to ordered micelles. *Physical Review Letters*, 96(12), Mar 31 2006.

[20] B. Capone, C. Pierleoni, J.-P. Hansen, and V. Krakoviack. Entropic Self-Assembly of Diblock Copolymers into Disordered and Ordered Micellar Phases. *Journal of Physical Chemistry B*, 113(12):3629–3638, Mar 26 2009.

[21] F. Eurich, A. Karatchentsev, J. Baschnagel, W. Dieterich, and P. Maass. Soft particle model for block copolymers. *Journal of Chemical Physics*, 127(13), Oct 7 2007.

[22] E. J. Sambriski and M. G. Guenza. Theoretical coarse-graining approach to bridge length scales in diblock copolymer liquids. *Physical Review E*, 76(5, Part 1), Nov 2007.

[23] I. W. Hamley. *The Physics of Block Copolymers*. Oxford University Press, 1999.

[24] F. S. Bates and G. H. Fredrickson. Block Copolymer Thermodynamics - Theory and Experiment. *Annual Review of Physical Chemistry*, 41:525–557, 1990.

[25] C. Loebbe. Imaging organic compound assemblies - oligomers, polymers and alkane derivatives in scanning force microscopy. *JPK Instruments Application Report*, http://www.jpk.com, 2008.

[26] R. Garcia and R. Perez. Dynamic atomic force microscopy methods. *Surface Science Reports*, 47(6-8):197–301, 2002.

[27] M. L. Huggins. Theory of solutions of high polymers. *Journal of the American Chemical Society*, 64:1712–1719, Jul-Dec 1942.

[28] P. J. Flory. Thermodynamics of high polymer solutions. *Journal of Chemical Physics*, 10(1):51–61, Jan 1942.

[29] A. Yu. Grosberg and A. R. Khokhlov. *Statistical Physics of Macromolecules*. American Institute of Physics Press, 1994.

[30] K. F. Freed and J. Dudowicz. Lattice cluster theory for pedestrians: The incompressible limit and the miscibility of polyolefin blends. *Macromolecules*, 31(19):6681–6690, Sep 22 1998.

[31] Z. G. Wang. Concentration fluctuation in binary polymer blends: chi parameter, spinodal and Ginzburg criterion. *Journal of Chemical Physics*, 117(1):481–500, Jul 1 2002.

[32] E. Helfand. Theory of Inhomogeneous Polymers - Fundamentals of Gaussian Random-Walk Model. *Journal of Chemical Physics*, 62(3):999–1005, 1975.

[33] M. W. Matsen. The standard Gaussian model for block copolymer melts. *Journal of Physics-Condensed Matter*, 14(2):R21–R47, Jan 21 2002.

[34] D. A. Hajduk, P. E. Harper, S. M. Gruner, C. C. Honeker, G. Kim, E. L. Thomas, and L. J. Fetters. The Gyroid - A New Equilibrium Morphology in Weakly Segregated Diblock Copolymers. *Macromolecules*, 27(15):4063–4075, Jul 18 1994.

[35] L. Leibler. Theory of Microphase Separation in Block Co-Polymers. *Macromolecules*, 13(6):1602–1617, 1980.

[36] S. A. Brazovskii. Phase-Transition of an Isotropic System to an Inhomogeneous State. *Zhurnal Eksperimentalnoi i Teoreticheskoi Fiziki*, 68(1):175–185, 1975.

[37] M. W. Matsen and F. S. Bates. Unifying weak- and strong-segregation block copolymer theories. *Macromolecules*, 29(4):1091–1098, Feb 12 1996.

[38] A. K. Khandpur, S. Forster, F. S. Bates, I. W. Hamley, A. J. Ryan, W. Bras, K. Almdal, and K. Mortensen. Polyisoprene-polystyrene diblock copolymer phase diagram near the order-disorder transition. *Macromolecules*, 28(26):8796–8806, Dec 18 1995.

[39] D. A. Hajduk, H. Takenouchi, M. A. Hillmyer, F. S. Bates, M. E. Vigild, and K. Almdal. Stability of the perforated layer (PL) phase in diblock copolymer melts. *Macromolecules*, 30(13):3788–3795, Jun 30 1997.

[40] M. W. Matsen, G. H. Griffiths, R. A. Wickham, and O. N. Vassiliev. Monte Carlo phase diagram for diblock copolymer melts. *Journal of Chemical Physics*, 124(2), Jan 14 2006.

[41] G. H. Fredrickson and E. Helfand. Fluctuation Effects in the Theory of Microphase Separation in Block Copolymers. *Journal of Chemical Physics*, 87(1):697–705, Jul 1 1987.

[42] S. Stepanow. Extension of the Theory of Microphase Separation in Block-Copolymer Melts Beyond the Random-Phase-Approximation. *Macromolecules*, 28(24):8233–8241, Nov 20 1995.

[43] M. Muller and G. D. Smith. Phase separation in binary mixtures containing polymers: A quantitative comparison of single-chain-in-mean-field simulations and computer simulations of the corresponding multichain systems. *Journal of Polymer Science Part B-Polymer Physics*, 43(8):934–958, Apr 15 2005.

[44] K. Ch. Daoulas, M. Mueller, J. J. de Pablo, P. F. Nealey, and G. D. Smith. Morphology of multi-component polymer systems: single chain in mean field simulation studies. *Soft Matter*, 2(7):573–583, 2006.

[45] A. A. Louis, P. G. Bolhuis, J. P. Hansen, and E. J. Meijer. Can polymer coils be modeled as "soft colloids"? *Physical Review Letters*, 85(12):2522–2525, Sep 18 2000.

[46] P. G. Bolhuis, A. A. Louis, J. P. Hansen, and E. J. Meijer. Accurate effective pair potentials for polymer solutions. *Journal of Chemical Physics*, 114(9):4296–4311, Mar 1 2001.

[47] J. Mazur, C. M. Guttman, and F. L. McCrackin. Monte carlo studies of self-interacting polymer chains with excluded volume. ii. shape of a chain. *Macromolecules*, 6:872–874, 1973.

[48] M. Murat, G. S. Grest, and K. Kremer. Statics and dynamics of symmetric diblock copolymers: A molecular dynamics study. *Macromolecules*, 32(3):595–609, Feb 9 1999.

[49] Q. Wang, P. F. Nealey, and J. J. de Pablo. Lamellar structures of symmetric diblock copolymers: Comparisons between lattice Monte Carlo simulations and self-consistent mean-field calculations. *Macromolecules*, 35(25):9563–9573, Dec 3 2002.

[50] T. P. Lodge. Block copolymers: Past successes and future challenges. *Macromolecular Chemistry and Physics*, 204(2):265–273, Feb 20 2003.

[51] J. Hoshen and R. Kopelman. Percolation and cluster distribution. 1. Cluster multiple labeling technique and critical concentration algorithm. *Physical review B*, 14(8):3438–3445, 1976.

[52] A. Al-Futaisi and T. W. Patzek. Extension of Hoshen-Kopelman algorithm to non-lattice environments. *Physica A-Statistical Mechanics and its Applications*, 321(3-4):665–678, Apr 15 2003.

[53] F. J. Martinez-Veracoechea and F. A. Escobedo. Lattice Monte Carlo simulations of the gyroid phase in monodisperse and bidisperse block copolymer systems. *Macromolecules*, 38(20):8522–8531, Oct 4 2005.

[54] V. A. Ivanov, W. Paul, and K. Binder. Finite chain length effects on the coil-globule transition of stiff-chain macromolecules: A Monte Carlo simulation. *Journal of Chemical Physics*, 109(13):5659–5669, Oct 1 1998.

[55] H. Minkowski. Volumen und Oberfläche. *Mathematische Annalen*, 57:447–495, 1903.

[56] K. Michielsen and H. De Raedt. Integral-geometry morphological image analysis. *Physics Reports-Review Section of Physics Letters*, 347(6):462–538, Jul 2001.

[57] E. L. Thomas, D. M. Anderson, C. S. Henkee, and D. Hoffman. Periodic Area-Minimizing Surfaces in Block Copolymers. *Nature*, 334(6183):598–601, Aug 18 1988.

[58] Y. Nishikawa, H. Jinnai, T. Koga, T. Hashimoto, and S. T. Hyde. Measurements of interfacial curvatures of bicontinuous structure from three-dimensional digital images. 1. A parallel surface method. *Langmuir*, 14(5):1242–1249, Mar 3 1998.

[59] C. Gross and W. Paul. A soft-quadrumer model for diblock copolymers. *Soft Matter*, 6(14):3273–3281, May 26 2010.

[60] N. Metropolis, A. W. Rosenbluth, M. N. Rosenbluth, A. H. Teller, and E. Teller. Equation of State Calculations by Fast Computing Machines. *Journal of Chemical Physics*, 21(6):1087–1092, 1953.

[61] R. W. Hockney and J. W. Eastwood. *Computer Simulations Using Particles*. McGraw-Hill, New York, 1981.

[62] M. P. Allen and D. J. Tildesley. *Computer simulation of liquids*. Clarendon Press, New York, NY, USA, 1989.

[63] F. J. Martinez-Veracoechea and F. A. Escobedo. Simulation of the gyroid phase in off-lattice models of pure diblock copolymer melts. *Journal of Chemical Physics*, 125(10), Sep 14 2006.

[64] U. Micka and K. Binder. Unusual Finite-Size Effects in the Monte-Carlo Simulation of Microphase Formation of Block-Copolymer Melts. *Macromolecular Theory and Simulations*, 4(3):419–447, May 1995.

[65] H. Fried and K. Binder. Non-Gaussian conformational behavior in diblock copolymer melts - is the RPA valid. *Europhysics Letters*, 16(3):237–242, Sep 14 1991.

[66] K. Almdal, J. H. Rosedale, F. S. Bates, G. D. Wignall, and G. H. Fredrickson. Gaussian-Coil to Stretched-Coil Transition in Block Copolymer Melts. *Physical Review Letters*, 65(9):1112–1115, Aug 27 1990.

[67] R. J. Low. Measuring order and biaxiality. *European Journal of Physics*, 23(2):111–117, Mar 2002.

[68] A. Saupe. Recent Results in Field of Liquid Crystals. *Angewandte Chemie-International Edition*, 7(2):97–&, 1968.

[69] D. Frenkel and B. Smit. *Understanding Molecular Simulation*. Academic Press, San Diego, second edition, 2002.

[70] D. Van der Spoel, E. Lindahl, B. Hess, G. Groenhof, A. E. Mark, and H. J. C. Berendsen. GROMACS: Fast, flexible, and free. *Journal of Computational Chemistry*, 26(16):1701–1718, Dec 2005.

[71] W. Paul, D. Y. Yoon, and G. D. Smith. An Optimized United Atom Model for Simulations of Polymethylene Melts. *Journal of Chemical Physics*, 103(4):1702–1709, Jul 22 1995.

[72] B. Hess, H. Bekker, H. J. C. Berendsen, and J. G. E. M. Fraaije. LINCS: A linear constraint solver for molecular simulations. *Journal of Computational Chemistry*, 18(12):1463–1472, Sep 1997.

[73] S. Nosé. A Molecular-Dynamics Method for Simulations in the Canonical Ensemble. *Molecular Physics*, 52(2):255–268, 1984.

[74] W. G. Hoover. Canonical Dynamics - Equilibrium Phase-Space Distributions. *Physical Review A*, 31(3):1695–1697, 1985.

[75] S. Maolin, Z. Fuchun, W. Guozhong, F. Haiping, W. Chunlei, C. Shimou, Z. Yi, and H. Jun. Ordering layers of [bmim][PF6] ionic liquid on graphite surfaces: molecular dynamics simulation. *J Chem Phys*, 128(13):134504, Apr 7 2008.

[76] W. Paul, G. D. Smith, and D. Y. Yoon. Static and dynamic properties of a n-C100H202 melt from molecular dynamics simulations. *Macromolecules*, 30(25):7772–7780, Dec 15 1997.

[77] K. Hur, R. G. Winkler, and D. Y. Yoon. Comparison of ring and linear polyethylene from molecular dynamics simulations. *Macromolecules*, 39(12):3975–3977, Jun 13 2006.

[78] V. A. Harmandaris, K. C. Daoulas, and V. G. Mavrantzas. Molecular dynamics simulation of a polymer melt/solid interface: Local dynamics and chain mobility in a thin film of polyethylene melt adsorbed on graphite. *Macromolecules*, 38(13):5796–5809, Jun 28 2005.

[79] K. C. Daoulas, V. A. Harmandaris, and V. G. Mavrantzas. Detailed atomistic simulation of a polymer melt/solid interface: Structure, density, and conformation of a thin film of polyethylene melt adsorbed on graphite. *Macromolecules*, 38(13):5780–5795, Jun 28 2005.

[80] A. K. Soper. Empirical potential Monte Carlo simulation of fluid structure. *Chemical Physics*, 202(2-3):295–306, Jan 15 1996.

[81] D. Reith, M. Putz, and F. Muller-Plathe. Deriving effective mesoscale potentials from atomistic simulations. *Journal of Computational Chemistry*, 24(13):1624–1636, Oct 2003.

[82] Q. Wang, Q. L. Yan, P. F. Nealey, and J. J. de Pablo. Monte Carlo simulations of diblock copolymer thin films confined between two homogeneous surfaces. *Journal of Chemical Physics*, 112(1):450–464, Jan 1 2000.

[83] E. Huang, T. P. Russell, C. Harrison, P. M. Chaikin, R. A. Register, C. J. Hawker, and J. Mays. Using surface active random copolymers to control the domain orientation in diblock copolymer thin films. *Macromolecules*, 31(22):7641–7650, Nov 3 1998.

[84] G. T. Pickett and A. C. Balazs. Equilibrium orientation of confined diblock copolymer films. *Macromolecules*, 30(10):3097–3103, May 19 1997.

[85] M. W. Matsen. Thin films of block copolymer. *Journal of Chemical Physics*, 106(18):7781–7791, May 8 1997.

[86] M. Kikuchi and K. Binder. Monte-Carlo Study of Thin-Films of the Symmetrical Diblock-Copolymer Melt. *Europhysics Letters*, 21(4):427–432, Feb 1 1993.

[87] M. Kikuchi and K. Binder. Microphase Separation in Thin-Films of the Symmetrical Diblock-Copolymer Melt. *Journal of Chemical Physics*, 101(4):3367–3377, Aug 15 1994.

[88] G. J. Kellogg, D. G. Walton, A. M. Mayes, P. Lambooy, T. P. Russell, P. D. Gallagher, and S. K. Satija. Observed surface energy effects in confined diblock copolymers. *Physical Review Letters*, 76(14):2503–2506, Apr 1 1996.

[89] D. van der Spoel, E. Lindahl, B. Hess, A. R. van Buuren, E. Apol, P. J. Meulenhoff, D. P. Tieleman, A. L. T. M. Sijbers, K. A. Feenstra, R. van Drunen, and H. J. C. Berendsen. *Gromacs User Manual version 3.3*. http://www.gromacs.org, 2005.

[90] R. W. Hockney, S. P. Goel, and J. W. Eastwood. Quiet high-resolution computer models of a plasma. *Journal of Computational Physics*, 14(2):148–158, 1974.

I want morebooks!

Buy your books fast and straightforward online - at one of world's fastest growing online book stores! Environmentally sound due to Print-on-Demand technologies.

Buy your books online at
www.morebooks.shop

Kaufen Sie Ihre Bücher schnell und unkompliziert online – auf einer der am schnellsten wachsenden Buchhandelsplattformen weltweit! Dank Print-On-Demand umwelt- und ressourcenschonend produziert.

Bücher schneller online kaufen
www.morebooks.shop

KS OmniScriptum Publishing
Brivibas gatve 197
LV-1039 Riga, Latvia
Telefax: +371 686 204 55

info@omniscriptum.com
www.omniscriptum.com

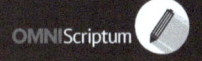

Printed by Books on Demand GmbH, Norderstedt / Germany